Cultural Evolution

'Philosophers of science do useful work when they rigorously analyze the conceptual usages and causal structure of a field of research. Tim Lewens' analysis of the debates over cultural evolution is very useful indeed.'

Peter J. Richerson, University of California Davis

'Contemporary human scientists are sorely in want of a theory of human evolution which, in breaking down taken-for-granted distinctions between biology and culture, nature and nurture, mind and body, emotion and cognition and so on, is able to bring together theorists across disciplines. Tim Lewens' philosophical contribution re-ignites the debate as to how this is to be achieved within and across disciplinary (and sub-disciplinary) boundaries in biology, anthropology, and psychology. His insightful reading of the dominant theoretical positions is bound to provoke new (even impassioned) discussion.'

Christina Toren, University of St Andrews

'Admirably clear, critical but constructive, both friends and foes of the evolutionary study of culture will find their ideas changed by this book. It is a model of how a philosopher can add value to a scientific debate.'

Paul E. Griffiths, University of Sydney

Cultural Evolution

Conceptual Challenges

Tim Lewens

OXFORD
UNIVERSITY PRESS

Great Clarendon Street, Oxford, OX2 6DP,
United Kingdom

Oxford University Press is a department of the University of Oxford.
It furthers the University's objective of excellence in research, scholarship,
and education by publishing worldwide. Oxford is a registered trade mark of
Oxford University Press in the UK and in certain other countries

© Tim Lewens 2015

The moral rights of the author have been asserted

First Edition published in 2015

Impression: 1

Published in the United States of America by Oxford University Press
198 Madison Avenue, New York, NY 10016, United States of America

British Library Cataloguing in Publication Data

Data available

Library of Congress Control Number: 2015932992

ISBN 978-0-19-967418-3

Printed and bound by
CPI Group (UK) Ltd, Croydon, CR0 4YY

For Jo, my sister

Contents

Acknowledgements

This book seeks to expose and to evaluate a set of conceptual disputes concerning what we might mean by human culture, and how we might go about accounting for it. These disputes have riven biological and social anthropologists. My particular focus concerns evolutionary approaches to cultural change, and to the genesis of our capacity for culture. The risk with a project of this kind is that despite—or more likely because of—a determination to treat the concerns of all parties with sympathy, one ends up making enemies on all sides and friends on none. Fortunately, I have been blessed with constructive advice from a remarkably generous and varied group of individuals, who have given encouragement and criticism in judicious measure.

My primary debt is to the wonderful group of researchers who have been working in Cambridge on the European Research Council project, 'A Science of Human Nature?'. Riana Betzler, Adrian Boutel, Andrew Buskell, Christopher Clarke, and Beth Hannon all read an early draft of the manuscript in full, and gave valuable advice. More generally I have profited from their work and their genial company. I owe special thanks to Beth, whose talents as a researcher, editor, and administrator are peerless. I am also indebted to the group of senior project advisors in Cambridge—Nicky Clayton, Tim Crane, Robert Foley, and Marilyn Strathern—who have offered guidance at crucial junctures. Ulinka Rublack provided important advice on the German Reformation, and Sian Lazar and Christos Lynteris have helped with various issues in social anthropology. Finally, the project has so far welcomed three distinguished visitors—Kim Sterelny, Christina Toren, and Kevin Laland—and they, too, have read large chunks of the work and offered insightful criticisms.

Outside of Cambridge, Alex Mesoudi, Cecilia Heyes, and two anonymous readers from Oxford University Press read the manuscript in its entirety, and were exceptionally kind in giving detailed feedback. For comments on chapters, talks, and shorter sections of the book I am grateful to Andreas de Block, Mathieu Charbonneau, Heidi Colleran, John Dupré, Peter Godfrey-Smith, Paul Griffiths, Philippe Huneman,

Fiona Jordan, Dan Sperber, Edouard Machery, Olivier Morin, Grant Ramsey, Pete Richerson, and Nick Shea. Ken Reisman shared with me the text from an unpublished presentation on cultural evolution and power, which had a significant influence over the material I present in Chapter 7 of this book. The remarks of Tim Ingold at an early conference associated with the project were also of enormous value.

Peter Momtchiloff at Oxford University Press has been efficient and supportive as an editor. I must also thank both the Department of History and Philosophy of Science at the University of Cambridge, and Clare College, for allowing me a year of research leave in 2013–14. My departmental colleagues—especially Anna Alexandrova, Jonathan Birch, Hasok Chang, John Forrester, Tamara Hug, Nick Jardine, Stephen John, Louisa Russell, Simon Schaffer, David Thompson, and Joeri Witteveen— have all contributed in important ways to the completion of this project. I have more recent debts to CRASSH—and especially to Simon Goldhill and Catherine Hurley—for providing a wonderful environment in which to finish this book. As always, I am grateful for the enthusiasm of my children Rose and Sam, and the wisdom of my wife Emma.

Some passages of this book draw on my earlier publications. Chapters 1, 2, and 9 borrow elements from my (2012) 'Cultural Evolution: Integration and Skepticism', in H. Kincaid (ed.), *The Oxford Handbook of Philosophy of Social Science*, Oxford: Oxford University Press: 458–80. Chapter 1 also contains sections based on my (2013) 'Evolutionary Epistemology', in M. Ruse (ed.), *The Cambridge Encyclopedia of Darwin and Evolutionary Thought*, Cambridge: Cambridge University Press: 451–60. Chapters 2 and 5 contain short passages adapted from my (2012) 'The Darwinian View of Culture', *Biology and Philosophy*, 27: 745–53. Chapter 3 is an edited version of my (2013) 'Cultural Information: Don't Ask, Don't Tell', in M. C. Galavotti (ed.), *New Directions in the Philosophy of Science*, Berlin: Springer: 369–82. Chapter 4 reproduces a section from my (2012) 'Human Nature: The Very Idea', *Philosophy and Technology*, 25: 459–74. Chapter 8 draws on my (2013) 'From Bricolage to Biobricks™: Synthetic Biology and Rational Design', *Studies in History and Philosophy of Biological and Biomedical Sciences*, 44: 641–8.

Last, I am exceptionally grateful to the European Research Council who funded this work under the European Union's Seventh Framework Programme (FP7/2007–2013)/ERC Grant agreement no 284123.

Introduction
Darwinism in Dispute

On 4 September 1862, Charles Darwin wrote to his friend Asa Gray (professor of botany at Harvard University), proudly recounting his young son's remarkable insight:

Progress of Education.—one of my little Boys, Horace, said to me, 'there are a terrible number of adders here; but if everyone killed as many as they could, they would sting less'.—I answered 'of course they would be fewer' Horace 'Of course, but I did not mean that; what I meant was, that the more timid adders, which run away & do not sting would be saved, & after a time none of the adders would sting'.—Natural selection!! (Darwin 1862)

Anecdotes like this encourage the thought that the explanatory centrepiece of Darwin's theorizing was so elementary his eleven-year-old son could not only understand it, he could formulate it independently. This image of natural selection as an idea of exceptional power and exceptional simplicity has been popular among recent commentators (e.g. Dennett 1996). Richard Dawkins, for example, has written:

Charles Darwin had a big idea, arguably the most powerful idea ever. And like all the best ideas it is beguilingly simple. In fact, it is so staggeringly elementary, so blindingly obvious that although others before him tinkered nearby, nobody thought to look for it in the right place. (Dawkins 2008)

It is not surprising that, convinced of the exceptional power of this simple idea, many have suggested it should be freed from the confines of biology, and applied to cultural change in humans and other animals. When, a few years ago, a special issue of the *Journal of Evolutionary Psychology* appeared under the title 'Why Aren't the Social Sciences Darwinian?', the editors could take for granted their readers' agreement that things would go much better for social scientists if they embraced

Darwinism, by adopting a theory of cultural change based on natural selection (Mesoudi et al. 2010).

At the same time as a growing number of thinkers, typically with backgrounds in biological anthropology, cognitive psychology, and evolutionary biology, have argued for a Darwinian turn in the social sciences, many thinkers from social anthropology and history have been just as vocal in dismissing this approach. Tim Ingold has for years aimed to bring biological and social anthropology together, and yet he is adamant that because, as he puts it, 'Neo-Darwinism is dead', the sort of approach recommended by Mesoudi and collaborators is fundamentally wrongheaded (2013: 1). Maurice Bloch remarks that 'Social and natural scientists have come to hate each other,' going so far as to claim they are 'repulsed by each other's style and mode of presentation' (2012: 1). Bloch has consistently sought reconciliation between these divided approaches to our species (e.g. Bloch 1998), and yet he, too, opposes recent efforts to use Darwin to broker a truce.

What is going on in these disputes over Darwinism in the social sciences? This book aims to understand what it means to take an evolutionary approach to cultural change, and then to understand why these approaches have sometimes been treated with suspicion. I will argue for the value of evolutionary thinking, but I will not dismiss the concerns of sceptics as driven by mere prejudice, confusion, or ignorance. On the contrary, the virtues claimed for evolutionary approaches are sometimes exaggerated. Confusions about what these approaches entail are propagated by their proponents, as well as by their detractors. By taking seriously the problems faced by evolutionary approaches to culture, we show how such approaches can be better formulated, where their most significant limitations lie, and how the tools of cultural evolutionary thinking might become more widely accepted.

The first task for this book is to survey the varied stances on cultural change that have attracted evolutionary labels. *Historical* approaches are the most general. They aim to describe the means by which human cognitive capacities have been shaped over time. *Selectionist* approaches instead stress the centrality of a form of Darwinian competition between ideas, techniques, values, and other apparently cultural traits. *Kinetic* theories of culture rest on a more general form of 'population thinking'; that is, on a commitment to explaining cultural patterns as the results of aggregated interactions between individual humans. This threefold taxonomy is developed in Chapter 1.

Chapter 2 assesses the relative strengths of the kinetic and selectionist approaches, and argues for the former over the latter. Some selectionist approaches can be dismissed for seeking unnecessarily close mappings between processes of cultural change and genetic evolution. We should be especially sceptical of memetics—a variety of cultural selectionism— which assumes that for culture to evolve we must be able to locate a set of cultural replicators. It is important we do not tie the fate of cultural selectionism to the fate of memetics; even so, many of the arguments used to defend non-memetic selectionist approaches are unconvincing. The most compelling considerations raised in their favour point to the insights to be had from using formal populational models to understand cultural change. The problem is that these arguments support kinetic approaches in general, rather than selectionist approaches in particular.

If we want to understand the primary conceptual challenges affecting evolutionary theories of culture, we need to understand the problems faced by kinetic approaches. This is the task for the central chapters of the book. If one asks cultural evolutionists what they mean by 'culture', they typically respond by saying that culture is information, albeit information acquired by learning from others (e.g. Richerson and Boyd 2005: 5, Mesoudi et al. 2006). In Chapter 3, I expose a series of problems with the more detailed accounts these theorists offer for exactly how we are to understand this notion of information, but I also argue for an untheorized, more pragmatic understanding of 'cultural information' as a term with two faces. On the one hand, it is simply a shorthand for the populational profile of mental states that cultural evolutionary models typically deal with. On the other hand, it acts as an open-ended heuristic prompt, encouraging investigators to document the resources that enable cultural reproduction. While these two faces for cultural information cause occasional confusions, I argue that the informational culture concept is fundamentally unproblematic.

A significant proportion of the most interesting, and most respected, work in the kinetic tradition explores models of gene/culture co-evolution. Cultural changes, it is said, can have knock-on effects on genetic evolution, as when the spread of dairying affects the subsequent genetic evolution of lactose tolerance (Holden and Mace 1997). These co-evolutionary models stress interactions between biological and cultural forces, but they have been criticized precisely on the grounds that such forces cannot be distinguished. The natural and the cultural, it is sometimes said, are not distinct factors affecting each other, but aspects of the very same developmental processes. For similar reasons, analyses of evolutionary change in terms of the interaction of cultural and genetic inheritance 'channels' have attracted criticism from a broad group of thinkers who complain that there is no way to pull developmental systems apart such that one might demarcate these allegedly separable channels (e.g. Griffiths and Gray 2001, inter alia). Chapters 4 and 5 defend gene/culture co-evolutionary modelling, along with the value of talking in terms of inheritance 'channels', at the same time as endorsing scepticism about robust distinctions between human nature and human culture. While it might appear that the use of co-evolutionary models presupposes the existence of clear distinctions between genetic and cultural traits, this appearance is illusory. The pathway towards this conclusion begins in Chapter 4, where I offer a sceptical treatment of recent efforts to make the human nature concept respectable. Instead, I argue that an exceptionally permissive 'libertine' conception is the only account of human nature that is biologically respectable. Chapter 5 then moves on to show that cultural evolutionary theory requires nothing more than this libertine account, and that its superficial reliance on a distinction between natural and cultural traits is merely an artefact of how some of its models are casually described.

Kinetic approaches to culture have no basic commitment to the importance of cultural selection, but they do have a basic commitment to the utility of idealized models. Chapter 6 addresses worries about how one might apply these sorts of models in the domain of culture. I focus on a pair of criticisms from Tim Ingold (2007). He argues that cultural models are inevitably circular, and he also complains that potent ethnographic data must be neutered if they are to be apt for representation in formal mathematical models. Ingold's criticisms fail when put in general terms, but they do point the way to a more serious set of worries, which only become clear when we look in detail at the sort of reasoning that is grounded in cultural modelling. These models are sometimes used in ways that are epistemically circular, and they are troubled by problems when data are borrowed from one domain with the aim of supporting hypotheses in another. These criticisms do not show that we should give up on cultural models, but they do contain suggestions for how to use them more persuasively.

Whether they use formal models or not, kinetic approaches assume that cultural patterns can be understood as the aggregated results of interactions between individuals (Richerson and Boyd 2005). Cultural evolutionism is reductionist if it claims that all cultural change is best understood in this way. Chapter 7 examines the limitations of this reductionist approach, and in so doing it addresses Richard Lewontin's long-standing concerns with the ability of cultural evolution to handle phenomena of power (e.g. Lewontin 2005). Kinetic approaches have no trouble understanding simple phenomena of power, but they struggle to capture the influence of complex organizations, and they are not the right tools to use when one wants to understand small-scale systems with networks of heterogeneous actors. The varied levels of enthusiasm for cultural evolutionary approaches across such domains as biological anthropology, social anthropology, and history reflect, in part, the suitability of kinetic approaches to the different sorts of phenomena these disciplines study.

In addition to a reliance on various forms of populational modelling, cultural evolutionary theories typically make use of a form of methodological adaptationism. More specifically, cultural evolutionists argue that human minds contain collections of cognitive traits, which have evolved as a result of natural selection answering evolutionary problems posed to our ancestors in the Pleistocene. Some schools of evolutionary thought lay stress on dispositions to learn from others in characteristic ways; alternative evolutionary schools instead focus on dispositions to handle particular types of information in characteristic ways. Many cultural evolutionists also argue that reflection on these past adaptive problems offers us a valuable form of insight when it comes to understanding how our minds work today. Chapter 8 exposes these adaptationist commitments, before moving on to ask if there is anything objectionable about them. I do not deny that there is benefit to be had from 'adaptive thinking' of this kind. I do, however, argue that these benefits are limited, and that adaptive thinking will only have heuristic bite if cultural evolutionary thinking is highly deferential to the best work in ethnography, developmental psychology, neuroscience, and so forth.

The final chapter of this book offers a constructive case study for how an evolutionary approach to culture can proceed in a way that eschews problematic distinctions exposed in earlier chapters between individual and social learning; between traits that we owe to nature and traits that we owe to culture; between that which is universal, biological, and evolved, and that which is local, cultural, and learned. The case study concerns the prospects for a cultural evolutionary approach to the emotions.

Emotional states are frequently understood these days to be embodied. I draw on ethnographic and psychological work to argue for the direct influence of culture on emotions of all kinds, including what Ekman has described as culturally universal 'basic' emotions (1992). I argue that cultural evolutionists' conceptions of culture should be expanded to include various embodied states, that these states can be influenced by forms of learning and enculturation, and that to acknowledge all of this need not bring us into conflict with the claim that similarities in patterns of emotional expression across cultures are the results of inheritance from common ancestors. This approach to the evolution of the emotions is emblematic of an eclectic synthesis, which I take to be a desirable goal for work in cultural evolution.

We have no good reason to think that the social sciences will be unified by the construction of a Darwinian synthesis that has a notion of cultural selection at its core. We do have good reason to think that evolutionary tools—especially the tools of phylogenetic analysis, and the tools of populational modelling—can make valuable contributions as elements of a broad synthetic project in the social sciences, which will also draw on neuroscience, developmental and cognitive psychology, physiology, geography, ethnography, history, and maybe even philosophy.

1

What is Cultural Evolutionary Theory?

1.1 Species of Evolution

We begin with a platitude. Cultural evolutionists hold that, in some sense, culture evolves. There are plenty of ways of understanding what is involved in evolution, and so there are plenty of ways of taking an evolutionary approach to culture. This chapter fashions a serviceable taxonomy of these varied approaches. I distinguish the historical approach, the selectionist approach, and the kinetic approach. In very rough terms, the historical approach tells us that culture evolves, but only in the bland sense that it changes over time in a manner that is typically gradual. The historical approach is evidently unobjectionable, but also exceptionally permissive. It needs to feature in our taxonomy, both as a foil to more outspoken stances, and because some very prominent evolutionary thinkers—most notably Charles Darwin—are best understood as historical theorists. The selectionist approach adds meat to these bare evolutionary bones, by proposing that cultural items such as ideas, tools, techniques, or practices compete in a Darwinian struggle. One might assume that all modern cultural evolutionists are selectionists, but we will shortly see that for many—whom I call kinetic theorists— these narrow issues of cultural selection have secondary importance compared with what is, for them, the broader and more valuable business of constructing explanatory models that show how human populations have changed over time under the influence of various forms of learning.

1.2 The Historical Approach

The historical approach to cultural evolution is prompted by a wholly unconstrained conception of evolution as historical change. On this view, to understand a system in evolutionary terms is to understand it historically, which in turn means understanding how various forces have transformed its earlier states into later ones. Species evolve, and cultures evolve, but on this view the Himalayas evolve too, as does the cosmos. As is very well known, Darwin indicated towards the end of the *Origin* that his ideas would have implications for broad areas of inquiry well beyond the traditional domains of natural history: 'In the distant future I see open fields for more important researches. Psychology will be based on a new foundation, that of the necessary acquirement of each mental power and capacity by gradation' (1859: 488). Darwin did not say that mental powers would inevitably be explained by appeal to natural selection. Instead, he made a bet that mental powers in humans and other animals would be explained by appeal to gradual historical processes.

Consider *The Descent of Man* (Darwin 1871). Part of this book is devoted to an explanation of how 'the moral sense'—our sense that some actions are morally right, others wrong—emerged in humans. Darwin also asks why it should be that we act in ways that are both unreflective and reasonably effective in promoting the general good. The answer he gives is gradualist: the refinement of the moral sense and the accumulation of moral knowledge are achieved by small increments over long periods of time. In some places this gradualist story focuses on natural selection acting on groups of humans to promote a sense of sympathy; that is, a capacity to feel injury to others as though it were injury to ourselves, and a consequent disposition to help others when they are in pain or distress. But as Darwin's account moves on, he tells of how the sense of sympathy became enlarged to encompass not only members of our immediate communities, but members of other tribes, other countries, other races, and even other species. Darwin also tells of how our tendencies to act in ways that promote the welfare of others have been refined through patient observation of the effects of our actions on others. Importantly, these elements of his account, especially when they explain why moral action that is unreflective should also be appropriate, often draw on Lamarckian forms of 'use-inheritance', and they sometimes focus solely on the ways in which publicly disseminated rules for proper conduct can influence behaviour, with no effort at all to conceive of such processes in terms related to natural selection (Lewens 2007).

In allying Darwin with the historical approach to cultural evolution, I do not mean to imply that Darwin never suggested that processes of selection might explain cultural change. In *The Descent of Man*, Darwin quoted the linguist Max Müller with approval:

A struggle for life is constantly going on amongst the words and grammatical forms in each language. The better, the shorter, the easier forms are constantly gaining the upper hand, and they owe their success to their own inherent value. (Darwin 1871: 60)

Here, Darwin tells us that selection occurs whenever there are entities that struggle for existence, regardless of what other properties they might have. As he put it, 'The survival or preservation of certain favoured words in the struggle for existence is natural selection' (ibid.). But while Darwin acknowledges here that a form of selection takes place between cultural entities—in this case words—he rarely invokes such forms of cultural selection when he offers detailed explanations for changes in our own species over time. Darwin's discussion of the emergence of the moral sense in humans—where cultural selection is absent—exemplifies his more general historical approach to cultural evolution.

The problem, of course, with the historical conception of cultural evolution is that while it might help us to categorize Darwin's cultural theorizing, it does little to make sense of modern debates. No modern student of culture will deny that cultures have histories; nor will they deny that the capacities that enable us to learn from each other, to be moved by each other, to cooperate and to interact with each other have histories. The historicity and contingency of human practices and institutions are commonly stressed themes among thinkers who are vehemently opposed to cultural evolution (e.g. Ingold 1995), and also among thinkers who have found little use for evolutionary approaches (e.g. Toren 2012). If we say that an evolutionary approach to human culture is any approach that offers a historical genealogy for cultural patterns, then we render any diachronic theorist of culture an evolutionary theorist, whether they would willingly accept the appellation or not. If we want to make sense of why 'cultural evolution' has attracted enthusiasts and critics, we need to refine our understanding of what an evolutionary approach might entail.

1.3 Selectionism and the Kinetic Theory

Cultural evolutionary theories of the selectionist and kinetic types have much in common. Both can be understood as reactions from within the community of evolutionary theorists to mainstream accounts of evolutionary theory itself. Textbook presentations often assume that evolutionary processes must work on genetically inherited variation (Mameli 2004). This may be an excusable pedagogical device, and perhaps it is sometimes a reasonable empirical assumption, but, understood as a general conceptual truth, it is flawed. Researchers steeped in the traditions of evolutionary biology, and familiar with its explanatory tools, have often pointed out that genes are not the only things passed from parents to offspring, and they are not the only resources that can explain why patterns of variation in one generation should be inherited in the next (Griffiths and Gray 1994, Jablonka and Lamb 1998, Avital and Jablonka 2000, Jablonka and Lamb 2005, Richerson and Boyd 2005, Danchin et al. 2011). In the human species (for example) skills, values, folk knowledge, technical scientific knowledge, linguistic expressions, and so forth can also be passed from parents to offspring by formal teaching, by imitation, and by other forms of learning. If learned skills, or moral values, make a difference to survival and reproduction, then natural selection can potentially promote their spread, even if genetic inheritance does not explain their transmission. What is more, skills and moral values are not only transmitted 'vertically' from parents to offspring. They can be passed from children to their friends, from teachers to children, or from influential role models to other adults. Children can influence their parents, as much as parents influence children. Such forms of transmission further complexify the ways in which a population's make-up can change over time, forcing us to take into account 'horizontal' or 'oblique' inheritance in addition to simple vertical inheritance. Most modern thinkers who explicitly describe themselves as cultural evolutionary theorists typically use these sorts of insights to argue that a complete account of human evolution needs modification if it is to encompass all the forces that have shaped our own species—and perhaps some cognitively sophisticated animal species—over time (Richerson and Boyd 2005). This diagnosis of a deficiency in simple evolutionary models leaves open the question of how these models should be enriched. Selectionism and the kinetic theory answer that question in different ways.

1.4 Cultural Selectionism

The selectionist approach, as I use that label here, holds that the conditions required for natural selection to act are present in the realm of culture, just as they are present in the realm of organic evolution. Since the question of what these conditions are is contentious (Godfrey-Smith 2009), there is considerable disagreement about what form a selectionist approach to culture should take.

More or less everyone thinks that selection requires that offspring resemble their parents. Some argue on this basis that natural selection also requires the existence of *replicators* (e.g. Hodgson and Knudsen 2010). The question of what replicators might be is also contentious, but for present purposes we can think of them in functional terms. Replicators are typically understood as entities whose role is to ensure resemblance between parents and offspring, and they fulfil this role because they make copies of themselves (cf. Hull 1988: 408). This notion of 'copying' should be understood in a relatively demanding way. To say that an entity is produced by a copying process means that the structure of one token item causally explains why a second token has the same structure. When strands of DNA multiply, a double-stranded piece of DNA is unzipped to produce two single strands. Complementary nucleotides are then attached, bit by bit, to the single strands, producing a pair of double-stranded daughter molecules whose structure matches the parent because of this template-matching process.

As Dan Sperber (2000) has stressed, we must not assume too quickly that all processes of 'reproduction' (understood loosely) are also instances of 'replication', understood in this more demanding sense. Sperber imagines a song that is played by one tape player, recorded by another, and then played again. This is a process of replication. But we can also imagine that a song is played, the song is recognized by a computer, the computer riffles through a database of songs, and the computer plays the same song. As in the first case, the same tune is played twice, and the second playing occurs because of the first, but the second song is not copied from the first. In this second case, the structure of the second song is not generated by a process that is informed by the structure of the first: instead, the second song was already part of the computer's repertoire. So while we might well say that a song has been reproduced here, we should not say it has been replicated.

In Chapter 2 we will look in more detail at the question of whether cultural inheritance is underpinned by the copying of replicators. For the moment, we should note that proponents of 'memetics' (e.g. Blackmore 2000) have been influenced by arguments from Dawkins (1989) and Hull (1988) in favour of the necessity of replicators for natural selection. They argue that mainstream evolutionary work teaches us that genes are the primary (or perhaps the only) replicators at work in organic evolution. Genes (so the story goes) meet our conditions for being replicators, because they have the function of ensuring that babies grow up to look like their parents, and they discharge this function via a form of template copying. Contrast this with a stable gravitational environment. Without such an environment, babies will not end up looking like their parents. But in spite of the role of the gravitational field in explaining generational resemblance, the gravitational field does not copy itself, it does not have the function of bringing about generational resemblance, and therefore it is not a replicator.

The memetic approach to culture looks to ground talk of cultural evolution by finding cultural replicators, which might in turn underpin cultural selection. Memes are these replicators, and they are typically (but not always) thought of as ideas, beliefs, or mental representations more generally. To use one of Richard Dawkins's favourite examples, religious beliefs seem to be the sorts of things that move from brain to brain, and their success is dependent on the sort of outward behaviour— whether communicative, devotional, or otherwise—that they provoke in the individuals whose minds house them. Religious beliefs, for the memeticist, are cultural analogues to genes. Like genes, the successes of these replicators are determined by their effects on phenotypes.

It is important to note that cultural evolutionists sometimes reject the meme concept (e.g. Boyd and Richerson 2000), and they sometimes express neutrality with respect to it (e.g. Mesoudi et al. 2004). They may deny that replicators are required for selection, they may simply ignore that question, or they may regard it as an open one. The selectionist view of culture can begin simply by noting that the cultural world, like the biological world, appears to be a domain where variation is abundant, where some form of reasonably faithful reproduction is at work, and where different cultural variants enjoy differential reproductive success as a result. A cultural selectionist, as I use that label, need assert neither the existence nor the importance of cultural replicators.

Informal appeals to selection in various cultural spaces have a long pedigree. We have already seen Darwin's own acknowledgement of the existence of a form of natural selection—and a form of struggle—between alternative forms of words in a language. Many years later Karl Popper argued, again in a largely informal way, and without any effort to tie his ideas to the detailed explanatory machinery of evolutionary theory, that his favoured image of scientific progress as a process of 'conjecture and refutation' could profitably be understood as a process of selection among scientific theories: 'The critical attitude might be described as the result of a conscious attempt to make our theories, our conjectures, suffer in our stead in the struggle for the survival of the fittest' (1962: 68). Donald Campbell then built on Popper's foundations, to argue that:

1. A blind-variation-and-selective-retention process is fundamental to all inductive achievements, to all genuine increases in knowledge, to all increases in fit of systems to their environment.
2. In such processes there are three essentials: (a) Mechanisms for introducing variation; (b) consistent selection processes; and (c) mechanisms for preserving and/or propagating the selected variations. (Campbell 1974: 421)

Campbell here focuses on the manner in which a selection process can increase the 'fit' between a system and its environment—the suitability of a theory to its subject matter, the suitability of a tool to its job, the suitability of an organism to its habitat—and he tells us that such selection processes require, somehow or other, that when a feature that promotes this 'fit' appears, it must be preserved. Campbell stresses the importance of cultural selection while leaving open the question of whether preservation in this sense requires the existence of a distinct class of replicators.

George Basalla's rationale for approaching the history of technology from the perspective of selection is reminiscent of Campbell's:

Because there is an excess of technological novelty and consequently not a close fit between invention and wants or needs, a process of selection must take place in which some innovations are developed and incorporated into a culture while others are rejected. Those that are chosen will be replicated, join the stream of made things, and serve as antecedents for a new generation of variant artefacts. Rejected novelties have little chance of influencing the future shape of the made world unless a deliberate effort is made to bring them back into the stream. (Basalla 1988: 135)

Basalla talks here about innovations being 'replicated', but he makes no effort to argue for the existence of technological replicators, in the sense of items whose function it is to preserve resemblances between artefactual parents and their offspring. When Basalla writes that some innovations are 'replicated', he simply means that by virtue of being favoured they also tend, somehow or other, to be retained in a cultural repertoire. Here, again, we have an example of non-memetic cultural selectionism.

These selectionist views are largely informal heuristic devices intended to prompt investigators to ask after the causes of cultural change and differentiation. Basalla, for example, sets himself the task of finding the technological ancestors that have been gradually modified or brought into combination to produce innovations that one might naïvely regard as wholly original. He aims to recover the pressures, often in the shape of transformed demands from users, that have favoured some forms of innovation over others.

Selection is also placed in the foreground when we come to more recent, more systematic, and more formalized evolutionary accounts of culture. Mesoudi et al. (2004) argue in favour of taking an evolutionary approach in the domain of culture by arguing that culture has 'key Darwinian properties'. What they mean by this is that culture instantiates the properties required for selection. In their view, these are 'variation, competition, inheritance, and the accumulation of successive cultural modifications over time' (2004: 1). This does not commit them, of course, to the further view that cultural selection is the only important factor affecting cultural change. Natural populations of plants and animals exhibit heritable variation in fitness; even so, population sizes are sometimes so small that forms of 'drift'—where the fittest variants are not the ones that enjoy the greatest reproductive success—can be significant. Likewise, it is possible to argue that cultural analogues to drift can play important roles even when the conditions for the action of selection are satisfied. Indeed, Mesoudi himself, alongside many other cultural evolutionary theorists, has explored the role of cultural 'drift' in some detail (e.g. Bentley et al. 2007, Mesoudi and Lycett 2009).

A cultural selectionist, as I understand that term, is not someone who says that cultural selection is the most important force affecting evolving cultural populations: instead, it is someone whose evolutionary approach to culture is justified on the grounds that cultural entities are engaged in a form of competitive struggle, of the sort that is usually thought to underlie all processes of selection.

While selection, in this sense, constitutes the bedrock of Mesoudi et al.'s (2004) evolutionary view of culture, replication does not: they point out—quite correctly—that Darwin's own conception of natural selection merely required that offspring resembled their parents, with no commitment to replicators as the means by which such resemblance is brought about. Of course, that leaves open the possibility that replicators might be required for selection all the same—Darwin may simply have failed to appreciate the commitments of his theory—and Mesoudi et al. (2004) consequently regard the question of the role of replicators in cultural selection as a matter to be left open to theoretical enquiry.

1.5 The Kinetic Theory

When Mesoudi et al. (2004) discuss what it is for a theory of culture to be 'evolutionary', they point out that 'evolution' need not be understood in Darwinian terms. They regard Darwinian evolutionary theories as sensible ones structured around a fundamental commitment to cultural selection, while non-Darwinian evolutionary theories are instead committed to a range of dubious and discredited assumptions about the intrinsically progressive nature of evolutionary processes, and about higher and lower forms of culture. They associate such non-Darwinian theories with early evolutionary approaches to anthropology due to Tylor and Morgan. We will return to the question of evolution and progress in Chapter 2. Here I want to establish that some very influential modern theorists of culture, who also think of themselves as evolutionists, nonetheless accord a comparatively minor role to cultural selection in their theorizing.

Many cultural evolutionists motivate their approach not from the selectionist perspective, but instead from what I call the kinetic perspective. Their theorizing is prompted by the basic observation that the human ability to learn from others makes a difference to the ways in which human populations change over time. Learning allows such populations to change very quickly, it allows them to adapt without the mortality levied by natural selection, it allows individuals to acquire valuable skills from non-kin, it potentially allows subgroups to be established with their own well-marked and idiosyncratic differences in spite of significant migration between these subgroups, and it potentially allows populations to be overwhelmed by absurd or dangerous fads. These theorists take a fundamentally pragmatic stance on these matters, advising that we update, modify, and augment standard evolutionary models of populational change to take account of these sorts of processes, in whatever way seems most appropriate to the explanatory task at hand (Mameli 2008a).

Richerson and Boyd, and the thinkers they have trained such as Joseph Henrich, are good examples of kinetic theorists. I cast them as kinetic theorists not because they deny that cultural change can be understood as a selective process, but because claims about selection are only of secondary importance to them, given their primary motivation for taking an evolutionary perspective on culture. For example, Richerson and Boyd wait until a long way into their (2005) overview of the cultural evolutionary project before suggesting that 'The logic of natural selection applies to culturally transmitted variation every bit as much as it applies to genetic variation' (2005: 76). In the opening pages of that book, where their basic methodological and explanatory assumptions are laid out, they affiliate their approach to a Darwinian tradition in a manner that does not place selection in the foreground, but which instead stresses the importance of *population thinking*: 'Modern biology is fundamentally Darwinian, because its explanations of evolution are rooted in population thinking . . . Population thinking is the core of the theory of culture we defend in this book' (2005: 5).

'Population thinking' has meant many things to many theorists (Lewens 2009a, Witteveen 2013). Richerson and Boyd begin with the thought that beliefs, skills, and values can differ in how appealing they are to others, they can differ with respect to how easy they are to learn, some can cause the individuals who possess them to be more likely to be noticed and emulated, and some can make an individual more likely to die young without much chance of influencing others. Regardless of whether we regard some or all of these processes as akin to those at work in the domain of biological evolution, we can ask a variety of questions about why individuals are disposed to be affected by others in the ways they are, and we can also ask a series of questions about what happens when individuals with characteristic learning dispositions are aggregated in populations. They summarize their whole approach with the remark that 'The heart of this book is an account of how the population-level consequences of imitation and teaching work' (2005: 6).

Darwin's thinking about adaptation, the emergence of species, and so forth, did indeed rest on the assumption that these large-scale patterns extending across time could be explained in terms of the aggregated effects of many small-scale events that occur in the lives of individual organisms. Boyd and Richerson's cultural theorizing also asks us to explain cultural phenomena in terms of the aggregated effects of small-scale events; in this case, instances of learning. The question of exactly what population thinking might consist in, and how it differs from other explanatory stances, will be addressed in detail in Chapter 7. For the moment we should note how hard it is to give an account of this aggregative form of population thinking that makes it *distinctively* Darwinian (Lewens 2009a, 2010a, Godfrey-Smith 2009). The approach is shared by the kinetic theory of gases, which also tells us that phenomena such as pressure and temperature are the outcomes of many summed interactions of smaller elements within a volume of gas, and that by adding up these interactions in statistically sophisticated ways we can understand the behaviour of the aggregate. There is no particular reason to call a populational approach 'Darwinian': we could just as well describe it as a kinetic theory of culture.

Richerson and Boyd have an additional reason to call themselves 'Darwinian', which lies in their methodological adaptationism. When we ask why we learn from each other in the ways we do, cultural evolutionists often recommend reflecting on how alternative learning dispositions might have augmented reproductive fitness in our ancestral environments. Although they sometimes distance themselves from mainstream evolutionary psychology, typically on the grounds that mainstream evolutionary psychologists do not give proper credit to the importance of social learning for our species, the basic method of much cultural evolutionary work often has plenty in common with evolutionary psychology. These affinities are expressed clearly by Henrich and Boyd:

Evolutionary psychology proposes that all brains, including human brains, consist of numerous domain-specific mental processors designed by natural selection to solve problems that were recurrent in the evolutionary past. Cultural transmission mechanisms represent a kind of special purpose adaptation constructed to selectively acquire information and behavior by observing other humans and inferring the mental states that give rise to their behavior. (Henrich and Boyd 1998: 217)

The populational aspect of their thinking involves asking how individuals with characteristic learning mechanisms combine to produce population-level effects. The adaptationist aspect of their thinking advises that we can understand learning mechanisms themselves as adaptations that have increased reproductive fitness, by virtue of the beneficial effects of such learning.

Richerson and Boyd's cultural evolutionism involves a commitment to population thinking, and in addition to this it focuses on understanding interactions between cultural change and more traditionally conceived processes of natural selection acting on genetic variation. This starting point means they need not think of cultural change as a selection process in its own right, and they certainly do not need to locate cultural replicators. Instead, they are free to construct populational models that allow them to explore the consequences of learning and teaching in whatever manner seems most tractable and suitable. Their models sometimes assume that individuals are poor imitators of others and they sometimes assume that one individual's belief is the product of observing many others: they are not shackled, that is, by a prior constraint to pin down a cultural entity that is similar to a gene, in the sense that it can be unambiguously assigned to a single parent, and that it is copied with high fidelity (see Enquist et al. 2010). As Richerson and Boyd put it: 'Natural selection-like processes are sometimes important, but processes that have no analogue in genetic evolution also play important roles' (2005: 6–7).

It is worth noting here that Richerson and Boyd offer two different (albeit compatible) accounts of what natural selection might amount to in the cultural realm. Rather rarely, they adopt an expansive notion, as in the remark quoted earlier: 'The logic of natural selection applies to culturally transmitted variation every bit as much as it applies to genetic variation' (2005: 76). Here, they think of cultural selection as a process that applies whenever there is some systematic explanation for the differential retention of cultural variants across a population. More usually, they have a far narrower understanding of what 'selection' involves, as we will now see.

Their rather curious conception of natural selection is best understood if one begins by thinking about genetic transmission. In 'standard' cases of sexual reproduction we characterize meiosis as unbiased—i.e. we expect alternative paired alleles to have equal chances of being transmitted to offspring. Where there is instead some form of bias—i.e. where paired alleles have unequal prospects for transmission—population geneticists do not usually describe the resulting change in gene frequencies as due to selection, but they might instead refer to segregation distortion or meiotic drive. (Some may prefer to think of these phenomena as intra-genomic selection, but this complication will not concern us here.) Natural selection, by contrast, occurs when various factors differentially affect the chances that individuals with alternative genes will survive or reproduce.

Boyd and Richerson transfer this taxonomy of evolutionary processes, with its distinction between transmission and selection, to the cultural realm. When alternative cultural variants have different chances of being transmitted by one individual to a learner—perhaps because the learner finds one variant more memorable, more useful, more attractive, and so forth—Boyd and Richerson do not describe these cases as instances of selection, but rather as instances of biased transmission. If attractiveness or ease of learning underpins differential transmission, then they class them as instances of 'content bias'. Cultural natural selection, on their view, instead occurs when varying properties of different individuals give them varying chances of influencing the cultural make-up of those who interact with them (2005: 68–79).

To illustrate the difference, imagine we know that a population is gradually changing from being primarily chopstick users to being pri-marily knife-and-fork users. If this is occurring because people find using knives and forks easier than using chopsticks, this is not cultural selec-tion but biased transmission. If the population changes because knife-and-fork users are more visible than chopstick users, and therefore more likely to influence others, this is cultural natural selection. Importantly, this means that, when interpreted in Richerson and Boyd's usual narrow sense of the term, cultural selection is not necessary for cultural adapta-tion. Suppose a succession of individuals in a cultural group consistently choose the spear design that they find most congenial, because they perceive that it best enables accurate throwing over long distances. The result is that spears are refined and improved over time. Such a cumulative adaptive trend might be mediated solely by content bias, that is, by a non-selective process.

Boyd and Richerson acknowledge that their understanding of which processes should be called 'selection' is somewhat idiosyncratic; however, the question of which processes it is appropriate to describe as 'selection-like' hardly matters to their overall theoretical project. Unlike memeticists, they have no prior commitment to claims about the importance of selection, or about its unique ability to explain cultural adaptation; as we have seen, they view biased transmission as a non-selective process that helps explain adaptation. Moreover, the value of Boyd and Richerson's models for thinking about how (for example) content bias or various other sorts of bias affect population change does not turn on which of these biases are properly described as 'selection-like'. An alternative way of making the same point is to imagine that we could somehow show conclusively that Boyd and Richerson's taxonomy of evolutionary processes was flawed, perhaps by drawing on analogies with intra-genomic selection, and that they should in fact count various forms of content bias as instances of selection. This would have no detrimental impact on the more general explanatory value of their theory and its models.

1.6 Cultural Epidemiology

Once we note Richerson and Boyd's fundamental commitment to population thinking, we also see that their approach has more in common with that of Dan Sperber and colleagues than one might initially think. Sperber has disagreed with meme theorists for assuming too easily that when beliefs reappear in a population with reasonable stability, this must be the result of faithful copying of underlying cultural replicators (Sperber 2000). Sperber and his co-authors have also expressed doubts about the accuracy and utility of thinking of cultural change as the result of a selection process (Claidière et al. 2014). For that reason, some commentators understandably think of Sperber's position as non-Darwinian. But Sperber's own 'epidemiological' approach is populationist through and through (Sperber 1996).

Sperber and Claidière say explicitly that they 'agree with Richerson and Boyd (2005) that Darwin-inspired population thinking provides the proper approach to the scientific study of cultural evolution' (2006: 22). Sperber, too, thinks of 'population thinking' as a framework that allows us to understand the aggregated effects of interactions between individuals:

All epidemiological approaches to cultures consider cultural phenomena as a population of mental or artifactual items distributed in a biological population (in particular a human population) and its habitat, and seek to explain the evolving distribution of these cultural items. Epidemiological approaches themselves are forms of 'population thinking' applied to cultural phenomena. (Sperber 2006: 439)

In more recent work, Claidière et al. stress a similar set of commitments. They describe their thinking as Darwinian, but that is because it is explicitly populational, rather than selectional. As they put it, 'Population thinking involves looking at a system (such as culture) as a population of relatively autonomous items of different types with the frequency of types changing over time' (2014: 2). What we might call the Sperber school, like that of Richerson and Boyd, is committed to an open-ended series of investigations that aims to explain why the frequencies of representations of different types change, and why they are maintained. They all espouse a kinetic theory of culture.

I wrote earlier that Richerson and Boyd combine this kinetic theory of culture with an adaptationist stance on the cognitive mechanisms that make culture possible. The same is the case for Sperber. Near the beginning of his important book *Explaining Culture*, Sperber spells out his own combination of an adaptationist stance on cognitive mechanisms with a populational stance on culture itself: 'The approach to culture advocated here is both epidemiological and cognitive, and, as will be seen, is more closely linked to Darwinism on the cognitive side than on the epidemiological side' (1996: 3).

Sperber's adaptationism is more overt than Richerson and Boyd's, and more strongly affiliated to Cosmides and Tooby's school of evolutionary psychology:

The brain is a complex organ. Its evolution has been determined by the environmental conditions that could enhance or hamper the chances of our ancestors to have offspring throughout phylogeny. There are good reasons to believe that the brain contains many sub-mechanisms, or 'modules', which evolved as adaptations to these environmental opportunities and challenges. (Sperber 1996: 113)

In summary, the schools of Sperber on the one hand, and Richerson and Boyd on the other, share a pair of very general commitments to a combination of population thinking and adaptationism.

1.7 Holes

What Sperber's school and the school of Richerson and Boyd have in common is an adaptationist, kinetic approach to culture that rejects stronger versions of replicator-based and selectionist approaches to cultural change. In saying this, I do not mean to suggest there are no significant differences between their approaches. They disagree regarding what sorts of cognitive adaptations in fact exist, and how these adaptations affect the ways in which the make-up of a population is conserved and altered. Some of these differences will be discussed in the remaining chapters of this book. What, then, is the value of a taxonomy that lumps them together, when a finer-grained taxonomy might have split them apart? The answer is that by understanding what these different pro-grammes have in common, we understand why both approaches have been dismissed by those sceptical of a populational approach to explaining cultural change (discussed in detail in Chapter 7), and by those sceptical of the adaptationism that underlies both programmes (discussed in detail in Chapter 8).

The taxonomy of evolutionary approaches to culture proposed here is not new. It corresponds well with earlier schemes proposed by myself (2009a, 2010a), Godfrey-Smith (2009, 2012), and Claidière et al. (2014). We have all stressed the (more or less) nested character of replicator-based, selectionist, kinetic (or 'populational'), and historical approaches. To illustrate the nested character of these approaches, consider that one might take a gradualist, historical stance on cultural change while asserting that the idiosyncratic behaviour of individual agents and organizations is often so significant that statistical aggregation is useless for understanding how cultural systems evolve. One might approach cultural change through the lens of aggregative mathematical models, while rarely crediting cultural selection with an important role. Last, one might regard cultural selection as ubiquitous, while denying that cultural reproduction is achieved by the action of replicators.

The taxonomy put forward here adds to these earlier, similar ones in a number of ways. First, it is useful to stress how the varied starting points of evolutionary investigation can push theorists in different directions. If one's primary interest is in formulating a general account of selection, then one is likely to seek commonalities between processes of cultural change and processes of biological change. If, instead, one's interest is the more pragmatic one of understanding how learning can make a differ-ence to genetic evolution, then there is no special need to search for commonalities in process, and one can instead start on the business of building explanatory models tailored to the circumstances. Second, the approaches I have outlined in this chapter are not perfectly nested, and the fact that they are not is itself illuminating. Kinetic theorists of culture typically rely on mathematical models to explore how many small events combine to produce broader cultural patterns. Cultural selectionists often use models of this sort, but not all cultural selectionists are kinetic theorists. Informal appeals to cultural selection, such as those typical of Basalla's work on the history of technology, are rarely accompanied by mathematical modelling, and they are just as likely to offer local narrative accounts of why one cultural variant prospered over competitors, as they are to offer images of broad cultural change as a result of many aggregated events.

The taxonomy offered here is imperfect, because there are a number of important themes in the theory of cultural evolution that it struggles to classify. Much of Kim Sterelny's recent work, for example, offers a narrative account of how our species became so good at accumulating and mobilizing large storehouses of valuable know-how (Sterelny 2012). His work is almost always historical, he rarely builds mathematical models (yet he often takes advantage of the model-based reasoning of others), and he often draws on the efficacy of forms of selection acting on cultural groups. His work is far from a paradigm of the kinetic approach, and yet it can hardly be written off as merely historical: it is perhaps best understood as a case of cultural natural history in the style of Darwin (1871). An even more obvious omission is cultural phylogenetics, an increasingly fruitful domain of enquiry (O'Brien and Lyman 2002, Gray et al. 2007, O'Brien et al. 2012). In addition to studying the mechanisms underlying various forms of cultural inheritance, and their impact on populational change, we can also attempt to reconstruct the patterns of cultural change. Mainstream biologists have developed powerful methods for uncovering the structure of evolutionary trees: that is, for understanding which species split from which others and when. It seems clear that cultural items of many kinds—most obviously languages, but also texts, tools, and techniques—also stand in recognizable genealogical relationships, and this has led many biological anthropologists to use phylogenetic methods from the biological sciences to reconstruct the history of borrowings in the cultural realm (Mace and Holden 2005, Mace and Jordan 2011).

Even if old and leaky, the taxonomy put forward in this chapter suffices for our voyage. It helps us to see that cultural evolutionary thinking should not be rejected simply because one might be sceptical of the importance, or even the conceptual coherence, of processes of cultural selection. Such thinking should also not be rejected on the grounds that one doubts that ideas, behaviours, or skills should be thought of as analogous to genes. Conversely, the taxonomy helps us to understand that opposition to cultural evolutionary theory is often driven not by scepticism about cultural selection, but by far more general scepticism regarding the kinetic theory of culture. Is it appropriate to think that cultures are made up of distinct particles, whose combined behaviour is best understood with tools of statistical aggregation? Is it reasonable to investigate the overall trajectory of a culture using simplified mathematical models? These questions, which affect all kinetic theories, constitute some of the most important challenges facing evolutionary approaches to culture. A significant proportion of this book is devoted to answering them. First, though, we need to understand better the relative merits of the selectionist and kinetic approaches. That is the job of Chapter 2.

2

The Kinetic Theory of Culture

2.1 Memes to Models

In Chapter 1 we distinguished three approaches to cultural evolutionary theory. The bulk of this book evaluates a series of challenges facing the kinetic approach, and in this chapter I explain why that will be the main focus. The basic reason is simple: the kinetic approach is more general than the selectionist approach, and it is arguably the approach that has shown the most promise. Kinetic theorists of culture, such as Boyd and Richerson, frequently build explanatory models that make no reference to cultural selection, and their approach explicitly denies that cultural selection, in their own strict sense of the term, is necessary for cultural adaptation. They do, however, share with selectionists a more general set of commitments to an image of social and cultural phenomena as the aggregated products of individual interactions, to a conception of culture in informational terms, to the value of mathematical modelling as a tool for explaining cultural phenomena, to an image of evolutionary change as resulting from interacting genetic and cultural inheritance systems, and to the heuristic value of a certain kind of adaptationist reasoning. Some of these generic commitments are also shared by those who are more explicitly opposed than Richerson and Boyd to selectionism, such as Sperber and collaborators (e.g. Claidière et al. 2014). The most penetrating criticisms of evolutionary approaches have focused on these generic commitments, rather than on any more specific commitment to cultural change understood in selectionist terms. We should prioritize them in our assessment of the promise of cultural evolutionary theory.

This chapter also casts sceptical doubt on the cases sometimes made for selectionism. I begin by discarding the stronger versions of selectionism that aim to discover cultural replicators. We should also be concerned by the manner in which weaker versions of selectionism can orient theorists to questions with little pragmatic pay-off, such as the issue of whether cultural change is Lamarckian or Darwinian. Finally, the case for selectionism is sometimes founded on a dubious historical inference from the successes of the modern synthesis within mainstream evolutionary biology.

2.2   Memes

As we saw in Chapter 1, meme theorists seek to understand cultural change using models which make the analogy between biological and cultural evolution very close indeed, and which draw on one rather specific account of biological evolution (Blackmore 2000, Dawkins 1989, Dennett 1996). Replicators, the story goes, are required for evolutionary processes to occur, because evolution requires some entity that can explain resemblance across generations. Roughly speaking, replicators discharge this function by making copies of themselves. The effects of replicators on the 'vehicles' that house them make a difference to the rate at which these replicators are copied. Evolution in general is a matter of the differential survival of replicators in virtue of these effects.

If genes are the paradigmatic replicators in the biological realm, memes are the replicators that underpin cultural evolution. In a famous list, Dawkins (1989: 192) gives some examples of putative memes. They are 'tunes, ideas, catch-phrases, clothes fashions, ways of making pots or of building arches'. What all of these things are supposed to have in common is that they are 'contagious'. The gist of the meme theory is easiest to appreciate if we focus on ideas. Different ideas—they might be scientific theories, moral values, or conceptions of the supernatural—spread from mind to mind. Alternative views about the nature of physical reality, the proper treatment of animals, or the character of God appear at moments in human history in the minds of just a few individuals, and can increase their representation in a population. They do so at different rates, and their abilities to make copies of themselves depend on their adaptive 'fit' with their local environments. These environments are cultural, and are partly constituted by the successes and failures of pre-existing memes. For example, the question of how likely it is that a given scientific hypothesis will take hold in a community of investigators depends, in part, on how well it integrates with what they already believe.

2.3 Passivity

All this talk of replication might lead one to think that memetics inevitably casts humans in a passive role. Blackmore, Dawkins, and Dennett, all enthusiastic proponents of the meme, sometimes encourage this impression with their talk of 'viruses of the mind', of the self as a 'pack of memes', and so forth. This may encourage an image of people as inadvertently colonized by ideas, technologies, and values, which they control no more than they control the bacteria in their guts. Tim Ingold complains about this forcefully:

There is, first of all, the question of how an approach to evolution couched in terms of the replication, transmission and distribution of 'cultural traits' can accommodate historical agency. Recall that cultural traits are supposed to adapt to their environment by means of humans, rather than humans adapting by means of their cultural knowledge and skills. In this topsy-turvy world, it seems, human beings are but the means by which traits propagate themselves in an environment. (Ingold 2007: 16)

One might object that scientific hypotheses, say, do not make copies of themselves autonomously. When I came to believe that the theory of evolution by natural selection was well founded, I was not invaded against my will by a parasite that burrowed into my head, propelled entirely by its own energies. I made efforts to read biological works, to understand what colleagues told me, to grasp the concepts involved, and to balance evidence in favour of the view.

Sensible memeticists will not be troubled by any of this. They, too, endorse the active role of the thinking organism in the copying of memes. Even in the case of genes, replication is not literally autonomous in the sense of being independent of background conditions, and it is not a problem for memetics if humans also have very active roles in enabling the copying of ideas. But we should not take seriously occasional claims by memetics' more zealous enthusiasts about the 'disappearance' of the self in the light of meme theory (Dennett 2001); nor should we come to the view that it is memes, rather than thinking, deliberating agents, that are in control of cultural change.

For the most part, models of cultural evolution—whether they embrace memes or reject them—aim to track the incidence of various beliefs, practices, and so forth in a population. This kind of analysis is wholly compatible with the thought that humans are active choosers, evaluators, and users of beliefs, practices, or whatever. Many proponents of these models are in the vanguard of the theory of niche construction, which stresses the mistakes of thinking that adaptation is always produced by environments shaping passive organisms via natural selection, which offers as a corrective the thought that adaptation is often achieved through the alteration of environments by active organisms, and which explicitly endorses the role of choice in the construction of environments (Laland et al. 2001, Mesoudi et al. 2007).

2.4 Replication and Attraction

Worries about passivity are not damaging to memetics. More serious problems for the approach arise when we note the sleight of hand involved in describing cultural change in terms of differential reproduction. It is one thing to claim, in a rather general sort of way, that different ideas may spread at different rates. It is another thing to say that they do so by virtue of a copying process.

To characterize some idea as a meme is not merely to say that it is a cultural unit that appears and reappears reasonably reliably throughout a population, in a manner that allows its frequency to be tracked using statistical tools borrowed from population genetics. Rather, it is to claim that it spreads through a strict process of replication. Strands of DNA count as replicators because the structure of a given DNA strand is causally responsible for the resembling structure of a daughter strand (Godfrey-Smith 2000). Can we say the same thing about ideas? Is the structure of a given idea causally responsible for the structure of a resembling daughter idea?

Replication is most likely to be at work in the cultural realm when believers are most actively engaged in efforts at focused inferential attention towards their companions. Sometimes individuals struggle to figure out precisely what someone else thinks, and they form similar views in virtue of that process. Dan Sperber—a prominent critic of memes—occasionally writes as though any process of cultural reproduction that involves inference cannot also involve replication, but that is to make the conditions required for replication too demanding. If I aim to discover exactly what an esteemed colleague believes about Renaissance anatomy, then, of course, I must make complex inferences as I talk to her and read her works (see also Boyer 2001: 40). But in spite of all these inferences it remains the case that it is because my colleague believes such-and-such that I also come to believe such-and-such. Sometimes, then, inference and replication are compatible. Even so, Sperber's underlying point is a good one: similarity of token beliefs, and therefore the spread of beliefs through a population, is not always achieved by replicative processes.

In some cases, large numbers of people may come to believe the same thing because they all witness similar events, not because their ideas are copied from each other. In hybrid cases one group, sharing a given set of ideas, may structure a cultural environment such that other individuals are highly likely to learn ideas of the same type from that environment. Dan Sperber and Scott Atran have rightly pointed out that in many cases when ideas are reproduced, a given idea triggers the formation of a similar one in the mind of another in virtue of a commonly shared background conceptual repertoire and constraining psychological biases, not because of a strict copying process (Atran 2001, Sperber 2000). Consider the spread of religious belief through a population. For this to be achieved by a strict process of replication, then each new believer would have to reconstruct their own faith by a process of careful imitation of the faith of an existing model believer, such that the properties of the original belief produce and explain the similar properties of the new belief. Maybe sometimes the process of religious conversion really does work like this, especially if charismatic evangelists are involved. But, to paraphrase Boyer's view (2001), it might simply be that the widespread appeal of notions of agency, power, and so forth have the result that when an individual is exposed to a handful of believers, that individual tends to develop a similar conception of a deity, without any need for close attention to precisely what those initial believers may have had in mind.

Sperber uses the notion of an 'attractor', in his earlier work at least (1996), to draw our attention to the ways in which the ubiquity of roughly similar forms of mental representation in certain populations can be explained by the fact that they are more or less reliable products of the typical inferential processes and background beliefs that are found in that population. Of course, even when cultural change is controlled by the action of such Sperberian attractors, we can document the rates of spread of different ideas through a population, but such a possibility shows the danger of conflating the specific commitments of memetic views with those of more general evolutionary modelling techniques.

In some cases it is unclear exactly where cultural evolutionists stand on these questions. So, for example, O'Brien et al. (2010: 3797) say that 'Cultural traits ... are replicators in their own right,' but it is doubtful that they support strong meme-like views of cultural replication. They justify their view by saying that cultural traits 'exist at various scales of inclusiveness and can exhibit considerable flexibility' (ibid.). They also remark that:

> cultural traits serve as units of replication in that they can be modified as part of an individual's cultural repertoire through processes such as recombination (new associations with other cultural traits), loss (forgetting) or partial alteration (incomplete learning, personal experience or forgetting select components) within an individual's mind. (ibid.)

These comments all rightly draw attention to the ways in which the cultural repertoires of individuals and populations can change over time, in manners that are analogous to the means by which the genetic composition of a population can change. But none of these observations is enough to show that Sperberian attraction has an insignificant role in explaining the reliable reproduction of cultural traits, so they also fail to show that cultural traits are, in O'Brien et al.'s own paraphrase of Hull, 'entities that pass on their structure directly through replication' (ibid.).

As we saw in Chapter 1, cultural evolutionists frequently begin their theorizing with the pragmatic aim of integrating various forms of learning into evolutionary models. That starting point is neutral regarding the propriety of the meme concept, but the end result of this pragmatic project can undermine the meme. Henrich and Boyd, for example, have built a mathematical model showing that cultural evolution does not require replicators (Henrich and Boyd 2002). For cultural evolution to be cumulative—that is, for the scientific or technical achievements of one moment in time to be built upon and improved—there must be reliable inheritance of cultural traits at the level of the population as a whole. They show that this can be achieved with error-prone copying—i.e. a failure of replication—when individuals learn from each other, so long as the nature of their errors can be compensated for in some other way. Their model establishes that compensation can be achieved through 'conformist bias'—a proposed tendency of individuals to preferentially adopt whatever the most prevalent view happens to be in a given population. None of this shows that cumulative evolution is usually the result of conformist bias acting to correct unreliable learning, only that it might sometimes be. We will examine in more detail what can, and cannot, be achieved with such models in Chapter 6. For the moment we should note that Henrich and Boyd's model constitutes a possibility proof for cumulative evolution without replication; hence it is a refutation of the claim that cumulative evolution must involve memes.

2.5 The Meme's-Eye View

Memeticists sometimes claim heuristic advantages for their approach, but these virtues are overrated. They remind us that the adoption of a given idea, practice, or whatever, need not be in the interests of its users; it need only be in the interests of the proliferating meme (Dennett 1996). We should not explain the proliferation of suicide, say, by asking how suicide benefits those who kill themselves. If the suicide meme is able to find some means of spreading swiftly through a population of thinkers, it will do so, whether it benefits its hosts or not. This is a weak defence for the memeticist, because there are plenty of mainstream views about cognition that remind us that people make decisions all the time that are not in their best interests. We do not need memetics to expose the widespread existence of various forms of irrationality, weakness of will, self-deception, false consciousness, subconsciously motivated action, and so forth. We do need to understand these diverse forms of thinking better, and it is not clear that memetics will help us to do this. The weakness of the memeticists' response is further compounded by the vacuity of the claim that the interests we should track are those of memes themselves. What this means, simply, is that the likely proliferation of an idea through some society can be represented as the 'fitness' of that idea; to ask 'What's in it for the meme?' is just another way of asking 'What might make a given meme fit in this population?' But once again, any enlightenment to be had from this insight is parasitic on a vast range of more specific, local, and contextually sensitive concerns about why some ideas are more likely to proliferate than others. Taking the meme's-eye perspective on scientific theory change, for example, is just another way of setting out on a familiar investigation of the sorts of factors that determine which theories get accepted in a given research community.

A much better response to these charges of vacuity can be found in Sterelny's cautious defence of the meme concept (Sterelny 2006).

He concedes that in many cases anything that can be explained by citing a meme's fitness can also be explained in terms of the features of human psychology which make some cultural items more memorable, learnable, or valuable than others. But Sterelny also makes a plausible case for thinking that at least in some instances the features that make technologies, say, especially easy to copy are not contingent on the precise features that human psychology happens to take in local contexts. He suggests that spears, for example, are the sorts of things whose intrinsic constitution makes them easy to reverse-engineer (that is to say, it is easy for a user or observer to figure out how they were made), valuable even when one fails to copy a spear perfectly, easy to make small improvements to, and useful to almost anyone even in their rudimentary versions. This all means that spears are the sorts of things that would be likely to be taken up and improved upon even if the fine-grained details of human needs and learning abilities had been different. For that reason, it is appropriate to focus on the properties of spears themselves, over the precise details of human psychology, when explaining the proliferation of spear technology. And that, in turn, establishes a strong explanatory role for the spear meme.

Sterelny advises restricting the meme concept to material artefacts. Spears themselves are memes for Sterelny, not ideas about how to make them. He succeeds in showing that some artefacts have features that make it likely that they will be adopted given a broad range of plausible psychologies. This is a good response to anyone who attacks the meme concept on the grounds that it is the fine-grained features of human psychology, rather than features of the objects taken up and adapted by human users, that always explain cultural change. But Sterelny's response acknowledges that the question of what makes some cultural item apt for spread and refinement is a matter of interaction between features of cultural items and aspects of human psychology, even when the relevant psychological features (and hence the fitnesses of the memes in question) may sometimes be preserved across fairly broad counterfactual variation. Sterelny's argument encourages attention to the explanatory roles of features of artefacts themselves; this is something which has been underlined by a variety of theorists (including social and cognitive anthropologists) who have drawn attention to artefacts as repositories of cultural tradition and knowledge even when they have not made specific use of the meme concept (Clark and Chalmers 1998, Henare et al. 2007, Mithen 2000).

2.6 Lamarckian versus Darwinian Approaches

It has frequently been said that while biological change is Darwinian, cultural change is Lamarckian (see Kronfeldner 2007 for numerous examples). It has also been said, again with considerable frequency, that the question 'Is cultural change Lamarckian?' is a bad one to ask (e.g. Hull 1988, many of the contributors to Ziman 2000, Lewens 2002a, Hodgson and Knudsen 2010). It is indeed a bad question, and once we understand why, we also further undermine replicator-based accounts of cultural change.

The assertion that cultural change is Lamarckian is perfectly comprehensible, and perfectly correct, if it means either that skills acquired by learning during the lifetime of a parent can be transmitted to its offspring, or if it means that the techniques assayed by human learners are not 'blind variations', but are typically informed by various educated guesses, prior knowledge, and so forth. The assertion that cultural change is Lamarckian is, nonetheless, regrettable. When put forward without qualification, it may lead us to ignore the important distinction between the claim that acquired variation can be inherited, and the claim that variation is directed. The assertion can also give the false impression that Darwin himself eschewed Lamarckian forms of explanation, and yet much of Darwin's work, especially when he focused on culture and the mind, appealed to use-inheritance (e.g. Darwin 1872). Finally, and most importantly, it is regrettable because it overlooks the desire of some cultural evolutionists to conceive of cultural selection as a process distinct from organic selection, which operates on ideas, techniques, or some other type of cultural entity.

In the biological realm, it is common these days to understand a denial of Lamarckian inheritance as an affirmation of the thought that while modifications to the germ line can influence phenotypes, modifications to phenotypes have no influence over the germ line. Interpreted in this way, the question of whether cultural change is Lamarckian or Darwinian presupposes that we can make sense of the notion that there is some cultural germ line. The problem is that it is very hard to see how we could identify such a thing, in large part because we cannot identify a distinct class of cultural replicators (Lewens 2004, Sperber and Claidière 2008).

To identify some entity as a replicator is to say, at a minimum, that it functions to bring about offspring/parent resemblance by acting in such a way that its structure is causally efficacious in bringing about a similar structure in a daughter replicator (Godfrey-Smith 2000). DNA molecules have this feature: the process of complementary base-pairing and the manner in which the DNA helix can unravel make it the case that daughter strands resemble parent strands because of the structure of those parent strands. This also explains why a reasonable test for being a replicator is a form of manipulation: if zygotic DNA is tampered with, then these alterations will tend to appear in the DNA of the resultant organism's offspring. If the nose of an adult is cosmetically enhanced, then that alteration will not appear in the individual's offspring. Hence, strands of zygotic DNA are typically regarded as replicators, while noses are not.

The problem in the cultural realm is that people can observe and attend to all manner of alterations to techniques, tools, and behaviours, and they can ensure that these alterations are repeated when these techniques are copied. This means that more or less any element of cultural reproductive cycles can pass the manipulation test for being a replicator (Wimsatt 1999, Lewens 2002a). We have already established, through a consideration of Sperber's worry that cultural reproduction sometimes proceeds without anything like replication, that on those occasions when cultural replication does occur, it is because observers are paying close attention to the structural properties of some cultural item, with the result that this structure is indeed causally responsible for a resembling structure in the reproduced object. This has the result that the question of which elements pass the manipulation test for replication can constantly shift, depending on exactly what observers happen to be paying attention to. If a pot's rim is accidentally chipped in such a way that it pours liquid without splashing, then users might notice this and reproduce it. If a dancer slips and the audience prefers the improvised moves that result, then this altered performance can be copied by observers. Under the right circumstances, the structures of artefacts and the structures of performances can be causally responsible for resembling structures in daughter artefacts or daughter performances.

It is curious that Hodgson and Knudsen (2010), in their defence of an evolutionary approach to economics, argue against the utility of discussing cultural change in terms of Darwinism versus Lamarckianism, while at the same time upholding the value of a replicator-centred approach. Discussions of Lamarckianism often presuppose, falsely, that we can discern a stable cultural germ line. That false presupposition also under-mines any effort to discern a stable class of cultural replicators among the various processes that make for stable cultural inheritance.

2.7 Progress

It is time to turn away from replicator-centred approaches, and towards more general selectionist approaches to cultural change. Cultural evolu-tionary theorists themselves often suspect that hostility to their views is best explained by simple ignorance on the part of the objectors. On some occasions this may indeed be the explanation (Perry and Mace 2010). Consider the example of Roger Smith, a historian much influenced by social anthropological accounts of the human species. He complains that 'modern evolutionary accounts of human origins continue to reflect the belief that there is an essential human nature, the nature all people share through their common root' (2007: 27). As we will see in Chapters 4 and 5, some evolutionists have indeed been tempted by talk of human nature, but Smith's comments do not do justice to the very broad range of conceptions of human nature, and of the sorts of forces that can affect it, which are found in modern evolutionary thinking (Laland and Brown 2002). Some champions of cultural evolutionary theory have rejected the concept of human nature altogether; others have argued at the very least for the malleability of human nature at the hands of cultural forces.

It may also be the case that plain ignorance explains why theories of cultural evolution are sometimes linked to dubious notions of progress, and denigrated on those grounds. The term 'evolution' may evoke images of higher and lower civilizations, and of a general tendency for societies to pass in sequence through these progressively more elaborate stages. Mesoudi et al. (2004) try to distance Darwin himself from this progressive conception of evolution, but it is clear that Darwin regarded natural selection as a process with a fairly reliable tendency (albeit not a guaranteed one) to produce increasingly complex forms of organization (Lewens 2007). Recall, for example, Darwin's belief that:

At some future period, not very distant as measured by centuries, the civilised races of man will almost certainly exterminate and replace throughout the world the savage races. At the same time the anthropomorphous apes, as Professor Schaaffhausen has remarked, will no doubt be exterminated. The break will then be rendered wider, for it will intervene between man in a more civilised state, as we may hope, even than the Caucasian, and some ape as low as a baboon, instead of as at present between the negro or Australian and the gorilla. (Darwin 1871: 201)

Early anthropologists who explicitly identified themselves as 'evolutionary' also shared a commitment to this view of progress (Layton 1997, 2010, Descola 2013: 72).

Darwin's reputation is hardly important for Mesoudi et al.'s (2004) more basic defence, which rightly stresses that modern conceptions of natural selection have no strong link to notions of progress. The fitter variant will tend to replace the less fit in a given population, but there is no sense in which fitness can be equated with moral or technical superiority, and even when the fitter variant replaces the less fit, mean fitness can still decrease (Sober 1993). Since the 'fitter than' relation is not transitive, there is no guarantee that iterated cycles of selection will result in a population whose members would outcompete the organisms which made up the population at the beginning of the process (Lewens 2007). What is more, many cultural evolutionists are sometimes interested in finding ways of explaining how learning can produce results that, from the perspective of biological fitness, do not lead to progress at all (e.g. Richerson and Boyd 2005: chapter 5).

What is puzzling is that many critics of selectionist approaches continue to frame their attacks in relation to progress, at the same time as demonstrating their knowledge of the modern evolutionary approach. Adam Kuper, for example, is fully aware that the modern conception of evolution has no intrinsic connotation of progress (2000a). Kuper even claims that most modern social anthropologists probably believe that, in general, cultural change has been progressive (2000b). We cannot explain Kuper's views away as products of simple ignorance. He is emphatic that 'Darwinism is, of course, utterly opposed in principle to any teleological way of thinking' and yet he immediately adds that 'faith in progress is probably one of the subliminal attractions of any "evolutionary" theory of culture' (2000a: 179). What Kuper means by this remark is unclear. Perhaps he is speculating, recklessly, that progressive notions are lurking in the minds of all cultural evolutionary theorists. More likely, and more charitably, he is expressing concern about the ways in which evolutionary theories are likely to be received by unreflective readers. In any case, the lengths to which most cultural evolutionists go to distance themselves from commitments to progress suggest at the very least that they approach such connotations in a responsible manner.

2.8 Vacuity

It is time to move on to more serious worries for the cultural selectionist. We have seen that a central challenge for the memetic approach is to show what sorts of explanatory insights it affords us that cannot be arrived at just as easily from more traditional perspectives on cultural change. The same worry applies to selectionist approaches more generally. That worry is especially acute when we consider informal, unquantified approaches to cultural selection. Consider this example, taken from Basalla:

Structural features that were necessary for marine craft were either useless or counterproductive on boats plying western rivers. These streams were relatively shallow and seldom, if ever, had large damaging waves. In case of a storm a steamboat was never far from shore. Sails were not needed on a steam-powered vessel nor could they have been used effectively within the narrow confines of a river . . . Because of the different conditions found on inland waters, the sea-going model of the eastern steamboat was transformed into the river steamboat in less than fifty years. (Basalla 1988: 89)

Basalla does not offer an explicitly selectionist explanation for the transformation of the eastern steamboat into the river steamboat. Yet it is easy to see what such an explanation would look like. After removal to a novel environment, variants on the eastern steamboat plan were quickly thrown up and selected, by virtue of their better fit with the new demands of western rivers. The only achievement here is to show that a well-understood phenomenon can be reframed in an evolutionary idiom. After all, the notion that the make-up of made objects can be predicted or explained in terms of the demands of the situations in which they are put to use is hardly revolutionary. One of the most significant challenges for selectionist views of cultural evolution is to show what sorts of novel explanatory or predictive resources are afforded by appealing to selective processes (Sober 1992, Lewens 2002a).

Suppose the selectionist is neutral with regard to the existence of cultural replicators, and chooses to address the evolutionary processes affecting a variety of cultural items such as ideas, tools, or techniques. The question of how likely a technique might be to proliferate in a population will depend on a very broad range of factors. They include, but are not limited to, the chance of its being discovered independently, the availability of material resources that assist in its being learned and performed, the difficulty of bodily movements required to execute it successfully, the number and accessibility of competent teachers, the degree to which it can be broken down into elements that can be practised independently of each other, the perceived utility it has, its integration with existing sets of techniques, and so forth. Which of these aspects should be incorporated into our estimation of the 'fitness' of the technique? And how is our understanding of the technique's passage through a population enhanced by our choice to summarize the elements that affect its success by means of a fitness value?

The kinetic theorist of culture does not need to find principled answers to these questions. She can proceed in a more piecemeal fashion, constructing populational models that are tailored to the demands of the explanatory problem that is being addressed. Such models may make no reference to cultural selection, or cultural fitness, at all. This is the case with some of Henrich and Boyd's models. In an example that will detain us at length in later chapters, they explain the emergence of 'conformist bias'—i.e. an exaggerated tendency to copy the majority—in terms of natural selection acting on genetic variation (Henrich and Boyd 1998). They aim to show that individuals who learn in a conformist manner are likely to do better than individuals who instead rely on learning from their environments, or who learn by imitating a randomly chosen member of the population. Their model takes account of the effects of learning, but not in a way that relies on a notion of Darwinian struggle, or selection, going on among cultural traits. Henrich and Boyd's (2002) model, which shows how population-level reproduction can be achieved in spite of error-prone learning between individuals, also makes no reference to cultural selection. Neither model can be accused of demonstrating something that was already obvious to all. In both cases, the explanatory pay-off of the evolutionary stance derives from the surprising nature of aggregation, made visible by the use of mathematical models. In neither case does the pay-off derive from a notion of cultural selection.

If populations as a whole are to build progressively more refined tools, techniques, values, and so forth, then somehow or other successful traits must be retained, in order that they might be further built upon in future generations (Lewens 2010a). One could try to bring cultural selection back into the foreground by insisting that whatever cultural processes contribute to the differential success of one practice over another should, by definition, be understood as contributors to cultural fitness, hence to cultural selection. The problem with this way of doing things is that one loses the sort of taxonomic precision that allows us to distinguish (for example) cultural selection (in Richerson and Boyd's restrictive sense) from the action of Sperber's 'attractors'. Moreover, in the mainstream of population genetics it is not typically the case that every factor that predictably increases the frequency of a genotype is understood as a contributor to selection: that is precisely why mutation, migration, and transmission bias are understood as alternatives to selection, in spite of the fact that they, too, can increase the frequency of genotypes (Lewens 2010b).

For these reasons I am sceptical of the pragmatic case Mesoudi et al. (2004) make for the selectionist approach to cultural evolution. They ask an important question: what is the pay-off of claiming that cultural change is 'Darwinian'? They argue that the Darwinian approach will act as a heuristic spur to future work as the strengths and weaknesses of various Darwinian analogies are explored empirically. We can agree with this, but it amounts to a weak form of justification. We will learn from cultural Darwinism, in part because we will understand where it fails. That sort of heuristic value can be claimed for very many broad theories of cultural change, including approaches from classical economics, complexity theory, Marxism, actor-network theory, and so forth.

More substantially, they suggest that a commitment to selectionism allows researchers to 'borrow sophisticated techniques originally developed for studying evolutionary change in biology to analyse cultural change' (Mesoudi et al. 2004: 9). The problem with this response is that while many leading evolutionary theorists of cultural change do indeed use models that are adapted from those used by population geneticists, these forms of adaptation are sometimes loose. Many important cultural evolutionary models make no reference to cultural selection, or even to the conditions of competitive struggle usually thought to be necessary for cultural selection. The pay-offs that have issued from the formal modelling of cultural change point to a kinetic approach in general, not to a selectionist approach in particular.

2.9 The Argument from History

Mesoudi et al. (2004) give a final response to their 'Why bother with Darwinism?' question, to which I now turn. They say that the Darwinian approach is likely to encourage a fruitful Darwinian synthesis in the social sciences. This is a theme that Mesoudi and collaborators have explored in various venues (e.g. Mesoudi et al. 2006, Mesoudi et al. 2010). They note an apparent lack of progress in the social sciences compared with the biological sciences. They tentatively diagnose this as due to the presence of an evolutionary synthesis in the biological sciences, and its absence in the social sciences. Were the social sciences to go evolutionary, then significant progress would follow.

Some prominent social anthropologists have agreed about the lack of progress in their field, but it is hard to imagine how one could devise a measure of progress that allows the biological and social sciences to be compared. Even if lack of progress is acknowledged, it is not clear that the lack of attention to evolutionary processes is to blame. Social anthropologists, confronted with the problem of explaining cultural change and cultural diversity, need to offer an account of one culture to readers from what is usually a different one. Necessarily, this requires that one grapple with questions of the adequacy of translation, of the tools one has for making one group's perspective available to another, and of the nature of explanation in this domain—whether, for example, one should seek to evoke how things look to them, whether instead one should give a neutral account based on objective principles, and whether an opposition of these two approaches makes sense. This reflection also requires fieldworkers to think about the likely practical impacts of their writings on the people they study. In sum, anthropologists have found it difficult not to run headlong into complex philosophical and political questions, which biologists have often been able to evade (Risjord 2007).

How do Mesoudi et al. understand what an evolutionary synthesis is? They claim that:

The synthetic framework provided by evolutionary theory...has successfully integrated several disparate disciplines into a coherent research program, evolutionary biology, and has the potential to do the same for the study of culture. Just as Darwin drew upon evidence from zoology, botany, geology, palaeontology, and physiology, this paper has incorporated findings from anthropology, psychology, sociology, linguistics, and history, with the hope of integrating these traditionally separate disciplines. (2004: 9)

They are quite right to say that Darwin's theorizing drew on an extraordinary range of disciplines. The question is whether the moral for the social sciences is simply that they should also draw on insights from a wide range of disciplines, including (say) economics, sociology, biology, psychology, linguistics, history, and literary studies. That would amount to an eclectic theory, and a synthetic theory, but in what sense would that theory be distinctively evolutionary?

As we have already seen, in Darwin's own thinking on (for example) the evolution of the moral sense in man, natural selection is at times in the forefront, and at times gives way to discussion of learning, the dissemination of public rules of conduct, the rational revision of moral teachings based on observed consequences, and so forth. Darwin proposes an evolutionary synthesis in the sense that he draws together numerous disciplines to provide a historical account of change in species. Darwin's synthesis does not, however, always place natural selection in the foreground, and natural selection is not always used to stitch together diverse disciplines, especially not when Darwin is discussing what we would now think of as human cultural evolution. One cannot use Darwin's own works to argue that the social sciences should become Darwinian, if what one means by this is that a social-scientific synthesis must have natural selection at its core.

This historical argument has been further developed in Mesoudi's more recent work, where he describes a united set of biological sciences which 'for decades have been unified under a single theoretical framework: Darwinian evolutionary theory' (2011: 22). Progress in these sciences 'would not have been possible in the absence of a common, unifying framework' (ibid.). By showing progress in biology to be attributable to a Darwinian synthesis, Mesoudi implies that a similar Darwinian synthesis might unite the social sciences and bring similar progress to them.

How should we evaluate this historical argument? If any biological science has made progress in the last sixty years, it is molecular biology. Molecular biologists have had little use for evolutionary work over this period, and much molecular biology continues to be practised largely independently of evolutionary concerns. There may well be good reason to regret this isolation (Morange 2010), but one can hardly use the history of recent biology to argue that all of biology has been united around a key Darwinian insight, nor can one say that a Darwinian stance is a necessary condition for progress in biology.

2.10 For Pragmatism and Eclecticism in Cultural Evolution

Darwin may serve as a good model for a broad synthesis in the social sciences, although not for the reasons Mesoudi imagines. When Darwin discussed cultural change, he merely made the case for the positive benefits of a natural historical standpoint: he did not try to claim that selection must lie at the heart of an account of cultural change. Likewise, it is enough for cultural evolutionists to argue (as do Richerson and Boyd, and as Mesoudi does in his more modest modes) that their various mathematical models bring a valuable set of tools to the sciences of human culture, without taking the further step of arguing that an underlying set of Darwinian concepts is likely to unify these sciences in a synthesis that parallels biology's own. A fruitful approach to culture is likely to require an eclectic synthesis that recognizes the kinetic approach as part of a broader effort to draw together insights from, at the very least, ethnography, psychology, neuroscience, and biology.

The kinetic approach offers the best response to those who doubt the explanatory pay-off of the cultural evolutionary stance. That should not be surprising, because kinetic theorists typically begin by asking pragmatically significant questions, and only then adapt evolutionary models in an effort to answer them. Some of these questions are of a very broad contrastive sort. Why, for example, is our own species able to acquire increasingly detailed folk knowledge regarding local flora and fauna, in a cumulative manner, while the ability of other species (including reasonably cognitively sophisticated species) to engage in this sort of cumulative cultural evolution seems so much more limited? An answer to this sort of question must examine the ways in which various different forms of learning need to be structured if skills and items of knowledge are to be preserved and refined across populations. Such a theory may tell us what it is about the human species that explains our uniquely developed capacity for constructing scientific theories. Of course, it does not follow from this that all disciplines will benefit from cultural evolutionary approaches. It does not follow, for example, that evolutionary insights will have much to say to sociologists of knowledge, intent on investigating the local factors that might explain the uptake of Darwinism in France; but the fact that some students of cultural change have little to learn from evolutionary theory does not show that the theory has nothing informative to offer to anyone.

Richerson and Boyd's reliance on 'population thinking' expresses the conviction that these important questions can be answered through the construction of aggregative models. They seek to understand the changing constitution of groups of people over time, as they are affected by genetic inheritance, natural selection, and also by various forms of learning. Exploring this demands a rather abstracted, rough-and-ready characterization of individual psychologies, albeit one informed by psychological research. This kinetic approach has the potential to provide novel insights, because it is not at all obvious how the combined interactions of individual learning psychologies might yield patterns of change or stasis at the populational level. Mathematical models, sometimes reasonably complex ones, are needed to explore these phenomena. It is a surprising result, for example, to learn of the circumstances under which populations can be stable enough to enable cumulative evolution, in spite of error-prone learning.

We are now in a better position to lay out the problems that will occupy us for the remainder of the book. It is reasonable to focus on the kinetic approach to culture, for this is the evolutionary approach likely to have the greatest pragmatic pay-off. I will argue that such an approach should feature as an element in a more general, eclectic synthesis in the social and evolutionary sciences. To secure this claim, we need to assess a series of challenges faced not by selectionist theories of culture in particular, but by kinetic theories in general. These include the propriety of the explicitly informational manner in which culture is understood by kinetic theorists, the risks of using mathematical models that draw on highly abstract characterizations of cognition, and the limitations of approaches that see cultural trends solely as the outcome of aggregated interactions between individuals.

Some readers may feel that the first, and perhaps the most important, conceptual questions that cultural evolutionists must answer concern what culture is supposed to be. My own view is that these are not, in fact, particularly troubling questions for cultural evolutionary theory, but it is worth getting them out of the way at this early stage. Chapter 3, then, assesses the conception of culture as information.

3

'Culture is Information'

3.1 Cultural Information

What, if anything, is culture? Cultural evolutionists have repeatedly told us that culture is information, albeit information with a certain provenance. What, then, is information, and is the informational conception of culture defensible? These are the questions addressed in this chapter.

In Alex Mesoudi's overview of the cultural evolutionary project he says that 'culture is information that is acquired from other individuals via social transmission mechanisms such as imitation, teaching or language' (2011: 2–3; see also Mesoudi et al. 2006: 331). Mesoudi is following the lead of Richerson and Boyd, who define culture as 'information capable of affecting individuals' behaviour that they acquire from other members of their species through teaching, imitation, and other forms of social learning' (Richerson and Boyd 2005: 5).

We should not suppose that only cultural evolutionists have been attracted to informational conceptions of culture. At the same time as expressing scepticism regarding the evolutionary frameworks of Richerson and Boyd (2005), Sperber (1996), and others, Maurice Bloch tells us that

What has been called culture is . . . a non-genetic, very long-term flow of information, in continual transformation, made possible by the fact that human beings are different from other animals because they can communicate to each other vast quantities of data, some of which they then may pass on to others. (Bloch 2012: 20)

The appeal of the informational conception of culture extends beyond the boundaries of cultural evolutionary theory.

How are we to assess these informational visions of culture? In this chapter, I argue for a 'don't ask, don't tell' approach to cultural information. When we *ask* cultural evolutionists what they mean by 'information', they give us answers that have obvious weaknesses. One might think, then, that the solution is for philosophers to *tell* cultural evolutionists what they ought to mean by cultural information, and more specifically that we should do so by a suitable adaptation of our best current theory of genetic information. Philosophers of biology have had plenty to say over the past fifteen years or so about the notion of genetic information, especially since John Maynard Smith's (2000) agenda-setting paper. They have noted the widespread use of various apparently semantic notions within molecular and evolutionary biology: triplets *code* for amino acids; nucleotides constitute an *alphabet*; the genome contains *information* regarding proper development. We now have various different options for how to understand this kind of talk: genes embody naturalized information (e.g. Shea 2007 inter alia); talk of genetic information exemplifies a preformationist fallacy (Oyama 1985); the language of genetic information is a pernicious metaphor (Griffiths 2001); or a fiction that is sometimes useful (Levy 2011).

I will argue in this chapter that the notion of cultural information as used by cultural evolutionists is not easily assimilated to teleofunctional notions of information that have been defended by the likes of Shea and others. Fortunately, we do not need a highly theorized account of cultural information. The notion is best understood as an open-ended heuristic prompt which encourages an examination of the ways in which bodies of behaviours, skills, beliefs, preferences, and norms are reproduced from one generation to the next.

3.2 Don't Ask

Four accounts of information, offered from within the cultural evolution community, all have flaws. Here I briefly examine each one in turn.

Richerson and Boyd (2005)

Richerson and Boyd appear to offer a definition of information in their *Not By Genes Alone*: 'By information we mean any kind of mental state, conscious or not, that is acquired or modified by social learning and affects behavior' (2005: 5). This looks like a stipulative definition, but this cannot be the right way to interpret what they are up to, for elsewhere they stress that 'some cultural information is stored in artifacts' (2005: 61). So whatever they mean by 'information', they do not mean to equate it with 'mental state' (Sperber and Claidière 2008). In linking information to mental states, they are merely expressing their empirical view that 'cultural is (mostly) information stored in human brains' (2005: 61). They also take the view that some of the information that constitutes culture may be stored outside human brains: they give the example of traditional pots. Richerson and Boyd have not offered an account of what information is, just a claim about where most of it is located.

Richerson and Boyd's concession that some small proportion of cultural information might be stored in traditional pots is tantalizing, because it admits of two very different interpretations. They might be suggesting that these pots feature symbolic representations or inscriptions, i.e. that they store information in the everyday sense that a book stores information. But there is an alternative reading: pots store information in the sense that the resources required for the successful fashioning of a new pot comprise not just an artificer's skills, but the material structure of pots themselves. This second reading is invited by the nature of the example. It would be odd to mention pots, rather than books, if the point one wishes to get across is that some artefacts feature intentionally encoded information. The second reading is also confirmed by the context of discussion, for Richerson and Boyd are addressing the question of whether information stored in people's heads suffices to explain the reproduction of significant practices and objects. If old pots are used as models, and new pots are copied from them, then craftsmen rely on the structure of these old pots—and not just on their own portable know-how—when they come to make new ones. This point holds regardless of whether pots have anything like intentionally encoded inscriptions or symbols on them.

If our reason for saying that pots store cultural information is grounded in their role in cultural reproduction, then how far does cultural information extend? Suppose, as seems likely, that people learn how to milk cows by interacting with cows, perhaps through guided practice in the presence of an adept. An udder is required, then, for the skill of milking to be regenerated in later generations. Just as the production of a new pot cannot take place without pots, so the production of the ability to milk a cow cannot take place without cows. If some cultural information is stored in pots, should we also say that some cultural information is stored in cows? In which case, might one reasonably regard many other elements of the physical and biotic environments, which have also been affected by generations of humans, and which affect the development of future generations, as repositories of cultural information too? Or is there some important asymmetry between the cow case and the pot case? Without further clarification of the nature of information, the extension of the evolutionary culture concept is unclear.

Boyd and Richerson (1985)

In their early work, Boyd and Richerson offer a more formal definition of information as 'something which has the property that energetically minor causes have energetically major effects' (1985: 35). This is intended as a wholly generic account of information: presumably it is meant to evoke intuitive examples whereby small changes to informational 'switches' (whether they are literally switches in a designed control system, or metaphorical 'genetic switches' in developmental pathways) have magnified downstream effects on the systems they influence.

Boyd and Richerson take this account of information from an article by Engelberg and Boyarsky, who explicitly have cybernetic control systems in mind when they write that 'informational networks are characterized by (1) mapping, and (2) low-energy causes giving rise to high-energy effects' (1979: 318). They form this view by reflecting on the fact that 'the governor of a steam engine can be said to "instruct" a valve to let in more or less steam', and they generalize by concluding that 'causal links which we intuitively consider to be informational all have this amplificatory characteristic'. However, there are plenty of cases of information-bearing relations where the energetic inequality is reversed. An instrument's display screen can carry information about solar flares: here, an energetically major cause on the surface of the Sun has an energetically minor effect in the laboratory.

As a general account of information, Engelberg and Boyarsky's picture will not do. It also will not do as a specific account of cultural information. In illustration of their concession that some small proportion of cultural information may reside outside people's heads, Richerson and Boyd claim that church architecture contains information regarding the rituals one should perform (2005: 61). Their own much earlier definition of information would suggest that this can only be the case if it also turns out that a low-energy cause here gives rise to a high-energy effect. If we pass briefly over the difficulties one might have in securing the claim that church architecture is a low-energy cause compared with the high-energy effect of performing a ritual, it seems that this account saddles the informational culture concept with an irrelevant epistemic hurdle. Surely it is not necessary to demonstrate this form of energetic inequality prior to claiming that some structure contains cultural information.

Hodgson and Knudsen (2010)

Hodgson and Knudsen's primary goal in their (2010) *Darwin's Conjecture* is to produce a generalized evolutionary theory suitable to all domains, from organic to cultural evolution. Their own particular area of interest is in developing an evolutionary economics. Unlike Boyd and Richerson, they take the view that evolution—including cultural evolution—invariably requires the existence of replicators, and they understand these replicators as bearers of information. In very rough terms, their hope is that we can assess the informational content—or 'complexity' as they sometimes put it—of a replicator by comparing its own state with the state of a hypothetical replicator that would maximize the fitness of the interactor that houses it in the actual replicator's environment. The actual environment of an actual replicator, in other words, specifies an ideal template for an optimal replicator, and any actual replicator is rich in information to the degree that it conforms to that optimal replicator. Depending on the sorts of evolutionary processes we might be interested in, these replicators might include genomes, or bodies of know-how.

There are many worries about these proposals. Here, I focus on their informal aspects. Hodgson and Knudsen's approach is impractical, for it demands that we assess the information content of an ideal genome (or a body of know-how) that may not exist in any individual, prior to assessing whether real genomes (or bodies of know-how) are in alignment with this ideal. Even if such problems of measurement could be solved, there are more theoretical worries. Why should we think that there is any single optimal specification for one of these replicators? Might there not be several distinct replicator specifications that maximize interactor fitness? If so, how are we to assess distance between some actual replicator and the ideal?

This sort of problem becomes especially acute in the case of cultural evolution, for here the instability and manipulability of the environment of adaptation itself are especially vivid. A population in any given setting may have the capacity not merely to adapt to meet the demands of the environment, but to alter that environment so that it presents altered demands (Lewontin 1983, Odling-Smee et al. 2003, Lewens 2004: chapter 4). If we require some 'ideal' specification for a cultural replicator before we can determine the informational content of an actual replicator, it seems we need some principled set of constraints on how that ideal replicator can be permitted to alter the environment in question. But Hodgson and Knudsen suggest no way of doing this, and I cannot see any way of achieving it.

Mesoudi (2011) and Ramsey (2013b)

Mesoudi does not back his informational conception of culture up with an explicit definition of information, but he does give a list of examples. Information, he says, is 'intended as a broad term to refer to what social scientists and lay people might call knowledge, beliefs, attitudes, norms, preferences, and skills, all of which may be acquired from other individuals via social transmission and consequently shared across social groups' (2011: 3). While Mesoudi is quite liberal in the sorts of things that can count as forms of information, he nonetheless stresses that culture is 'information rather than behaviour' (ibid.). Grant Ramsey has also denied that culture should be thought of as behaviour, on the grounds that 'if culture is behavior, then culture cannot cause or explain behavior' (2013b: 460). Mesoudi's reasons for excluding behaviour from the definition of culture are similar to Ramsey's—he is concerned about avoiding circular forms of explanation—but this form of argument fails, because it proves too much (Lewens 2012d).

As we have seen, Mesoudi takes it that 'information' names a variety of states that include 'knowledge' and 'skill', while excluding behaviour. His insistence that if culture comprises behaviour, then it cannot explain behaviour, would appear to have the consequence that culture cannot explain knowledge and skill; and yet, cultural evolutionists typically require that culture can explain such things. Ramsey recognizes the logical strength of the circularity argument, and he consequently bites the bullet, requiring not merely that culture should not be defined in terms of behaviour, but that it should not be defined in terms of any cognitive states, including belief and knowledge. He therefore requires some notion of culture as information that puts conceptual distance between informational states and cognitive states. But the problem of circularity detected by Ramsey and Mesoudi is really no problem at all: even if 'culture' names a variety of cognitive and behavioural states, culture can still explain such states on the grounds that the cognitive and behavioural endowments of one generation (i.e. its culture) can contribute to the production of similar cognitive and behavioural endowments in a later generation. This is precisely the reproductive role that cultural evolutionists believe can be mediated by, among other things, social learning.

Mesoudi adds to these worries about circularity the thought that 'there are other causes of behaviour besides culture' (2011: 4). What he means by this is that a given form of behaviour might be innate, it might be caused by individual learning, or it might be caused by social learning. If, then, our definition of culture is to make room for alternative causes of behaviour, Mesoudi concludes that we had better not make behaviour a subspecies of culture. Again, though, this objection proves too much: it would also have the consequence that since knowledge and skills—or, for that matter, information in the abstract—can potentially be produced by processes such as individual learning, we had better not equate culture with knowledge, skills, or even with information more generally. Yet these are among the equations Mesoudi and Ramsey wish to make.

3.3 Don't Tell

At this point one might think it useful to look to philosophical theories to tell cultural evolutionists what they ought to mean by the notion of 'cultural information'. There are many options one could turn to, but here I focus on Nick Shea's 'infotel' theory (e.g. Shea 2007, 2012, 2013), on the grounds that is probably the most developed effort to understand the informational content of the diverse contributors to organic inheritance. Shea argues that genes can be said to contain information because it is the evolved function of the genome to bring about heritable variation. Shea's theory builds on, and complexifies, earlier teleosemantic accounts of genetic information due to Maynard Smith (2000) and Sterelny et al. (1996). His theory allows us to say, in a full-blooded manner, that genes represent environmental conditions, that they can be misread, that they contain instructions, and so forth. Shea's theory is not committed to objectionable forms of genetic determinism. It allows that many developmental resources interact to explain development and reproduction, while only some of those resources contain information about developmental outcomes. To give just one example, a gravitational field of suitable strength is required for offspring to resemble their parents, but there has been no selective history whereby the properties of the gravitational field have been altered in such a way that improves its ability to stabilize parent/offspring resemblance. On the other hand, the existence of forms of proofreading machinery, and so forth, do provide strong evidence that the functional role of the genome is to bring about these stable resemblances.

Shea's overall theory has significant heuristic pay-offs: it is undeniably useful to ask, for any process whereby offspring come to resemble parents, or more generally where one generation comes to resemble another, whether the process has the sorts of features that appear to indicate complex adaptation as an inheritance system. In Chapter 5 we will see that this kind of framework helps us to mobilize a series of important general reflections regarding the costs and benefits of general features of inheritance for adaptive evolution (cf. Jablonka and Lamb 2005). That said, even as we acknowledge the heuristic benefits of asking after the functional constraints on inheritance systems, we should deny that Shea puts his finger on the way in which 'information' is used in cultural evolutionary theory. This is not a flaw in Shea's theory, for he does not aim to offer an account of what all evolutionists mean when they speak of 'information': he aims instead to construct a useful notion of information, suitable for various philosophical and scientific projects.

A couple of brief examples from cultural evolutionary theorizing suggest it may be unwise to apply Shea's theory in this domain. Kim Sterelny's very recent work argues that skills in an offspring generation may sometimes be acquired because parents (i) engage in skilled activities, (ii) their engagement in these activities results in various tools, raw materials, and so forth lying around for experimentation by others, and (iii) offspring therefore get to experiment with these pre-prepared objects in ways that make their own acquisition of the relevant skills easier (Sterelny 2012). This is just the sort of set-up that one may wish to describe—and that Sterelny does describe—in terms of the flow of information from parental generation to offspring generation. The problem, however, is that Shea's appeal to evolved functions in determining what counts as an inheritance system, and his definition of informational states in terms of inheritance systems, together have the result that *prior* to the advent of specializing adaptations that improve inheritance, it is strictly inappropriate to speak of cultural information at all. And yet, Sterelny wishes to stress how valuable transmission of cultural information can occur without such specialization, and he wishes to point out that these forms of highly disaggregated information transfer may have been important in the early stages of our cognitive development.

Sterelny's story for human evolutionary change is one in which we become better and better adapted to making use of cultural information. The sort of story he has in mind begins with 'accidental' information transmission, whereby juveniles simply hang around with their parents, and the result is that they take advantage of an environment that is structured in such a way that they have both learning opportunities and suitable materials with which to learn. This felicitous structuring has been produced by the social action of a previous generation. Subsequent adaptive steps may include greater tolerance of experimenting juveniles, and eventually something like explicit apprenticeship. We do not need a teleofunctional account of information to harness the heuristic benefits of attending, as Sterelny does, to the ways in which the design features of cultural inheritance systems may improve over time (Lewens 2014).

We can retain something valuable from Shea's account, by stressing the kinship between his view and metaphor theories of genetic and cultural information (e.g. Levy 2011). Shea has a nice way of expressing the motivating intuition behind his theory of information (Shea 2012). The bills of purple-throated carib hummingbirds develop to match the flowers they feed from. This 'good match' does not derive from causal interactions with the flowers in question during development. And so we might ask, 'how does the developing hummingbird "know" what shape of bill to produce, to match available nectar sources in its local ecology?' (Shea 2012: 2237). Shea's idea is that on some occasions, this 'knowledge' or 'information' is contained in the hummingbird genome, and generated through cycles of selection across generations.

Informational talk is indeed encouraged when one asks, metaphorically speaking, 'How does this organism know how to behave?', or equivalently 'Where does the information come from?' In a widely cited paper by Danchin et al. (2004) on the impact of 'public information' on cultural evolution, the authors make use of a framework in which various forms of 'non-genetic information' are discussed. The only definition they give of 'information' more generally is extremely cursory: it is defined as 'anything that reduces uncertainty' (2004: 487). Of course, this alludes to a well-known account of Shannon information. But Danchin et al. do not literally mean to suggest that the uptake of information reduces the cognitive uncertainty of the animals they discuss. Information helps resolve uncertainty in the organism, in the sense that it causes some outcome that is appropriate in the circumstances. For that reason, one might just as well talk of a plant's uncertainty regarding how to develop being reduced by the presence of an environmental correlate of impending drought.

Importantly, Danchin et al. distinguish *signals*—'traits specifically designed by selection to convey information' (2004: 487)—from 'inadvertent social information'. In these latter cases, the behaviour of organisms leaves a trail of relevant clues, which can be used by conspecifics even though such behaviours are *not* modified by selection for these functions of transmission. Danchin et al. give several examples, including the ways in which an animal's choice of habitat or mating site might be affected by its encounters with the behaviour of others, or with the downstream effects of another's behaviour. These clues help to inform the organism regarding how to act, or how to develop. In some cases, then, behaviours are described as carrying information, in spite of the fact that these behaviours do not have the teleofunction of carrying information.

Any factor that 'reduces uncertainty' might be considered a source of 'information' feeding into the broad process of appropriate development. There are a number of potential information sources: suitable developmental responses may come from configurations of the genome, from explicit instruction, or from imitation of another. Another potential explanation, suggested in Sterelny's discussion of the acquisition of skills by juveniles, adverts to interaction with an environment that is already well structured for the generation of the skill in question. In all cases, talk of information becomes attractive whenever development, or behaviour, can be interpreted in such a way that it is guided in a suitable manner by the presence or absence of material factors, where these might include the configuration of genomic elements, the placement of raw materials for the construction of an axe, the mating behaviours of conspecifics, and so forth. These reflections are all supportive of Jablonka's account of information:

A source—an entity or a process—can be said to have information when a receiver system reacts to this source in a special way. The reaction of the receiver to the source has to be such that the reaction can actually or potentially change the state of the receiver in a (usually) functional manner. (2002: 582)

Shea's metaphorical question—'How does the organism know what to do?'—need not be answered by appeal to a structure that has been shaped by selection in order to facilitate resemblance with respect to adaptive characters, even though sometimes it will be answered in this way. I suggest, then, that searches for sources of cultural information are typically unencumbered by requirements that the structures bearing information have teleofunctions of inheritance (see also Levy 2011).

The account offered here departs from Shea's view, and it is only partially compatible with some widely discussed comments on information by Bergstrom and Rosvall (2011). Bergstrom and Rosvall focus on a notion of information, according to which information is something transmitted from one generation to another. They appear to require that information-bearing structures have naturalized teleofunctions of transmission. As they put it, 'Like naturalized views of semantics, the transmission notion of information rests upon function: to say that X carries information, we require that the function of X be to reduce uncertainty on the part of a receiver' (2011: 169). A little earlier in the article the teleofunctional requirement is even more explicit: 'Our aim with the transmission sense of information is [. . .] to identify those components of biological systems that have information storage, transmission, and retrieval as their primary function' (2011: 167). They ask:

So must we impose our own notions of what makes an appropriate reference frame in order to single out certain components of the developmental matrix as signal and others as noise? If we want to know how the information necessary for life was compiled by natural selection, the answer is no. In this case, we are not the ones who pick the reference frame, natural selection is. (2011: 168)

This passage begs the important question of whether the information necessary for life is indeed always compiled by natural selection. Why could it not be the case that some information necessary for life is compiled in some other way, perhaps by the intentional structuring of a learning environment, but perhaps as the accidental by-product of artisanal activities, or mating activities (cf. Jablonka 2002)?

Earlier in their article, Bergstrom and Rosvall remind us of a more general account of information: 'An object X conveys information if the function of X is to reduce, by virtue of its sequence properties, uncertainty on the part of an agent who observes X' (2011: 165). If we make use of this definition, we can dispense with the teleofunctional condition on information, and instead construe the function of an object contributing to development via a covert assumption that we should regard the organism *as if* it were an observing agent making use of the various resources to hand as it develops. Anything used by the organism for this purpose would then acquire the function of reducing the organism's uncertainty. There is even a suggestion that Bergstrom and Rosvall have something like this in mind themselves: consider their comment that 'A single individual can only look at its own genome and see a sequence of base pairs. This sequence of base pairs is what is transmitted; it is what has the function of reducing uncertainty on the part of the agent who observes it' (2011: 169–70). Of course, an organism does not literally look at its own genome, but its genome, just like various aspects of its social and technical environments, can affect its developmental trajectory in an adaptive manner. It will then be entirely legitimate to regard Danchin et al.'s 'inadvertent social information', or Sterelny's structured learning environments, as bona fide loci of information too, in spite of the fact that their own structurings have not been selected for the function of transmission. Moreover, we can then move on to ask, of any of these informational loci, the sorts of important heuristic questions Bergstrom and Rosvall recommend to us regarding channel capacities, redundancy, and so forth. We can also ask the sorts of questions recommended by Maynard Smith and Szathmary (1995), and Jablonka and Szathmary (1995), regarding the comparative benefits of different informational channels, and we will see in Chapter 5 that much valuable work in cultural evolutionary theory has taken this perspective, whereby the costs and benefits of different forms of inheritance are assessed (e.g. Jablonka and Lamb 2005). The heuristic functions of Bergstrom and Rosvall's 'transmission sense' of information do not require a teleofunctional underpinning.

3.4 The Genealogy of Cultural Information

The historical contexts in which information talk found its way into genetic and cultural evolution were very different. S. Knudsen (not to be confused with T. Knudsen, whom we met in Section 3.2) has argued that in the early years of research in molecular biology, during the 1940s and

1950s, semantic metaphors played an important heuristic role in formulating plausible hypotheses for the relationship between chromosomal structure and developmental pathways (Knudsen 2005). Schrödinger, for example, wrote that:

> In calling the structure of the chromosome fibres a code-script we mean that the all-penetrating mind, once conceived by Laplace, to which every causal connection lay open, could tell from their structure whether the egg would develop, under suitable conditions, into a black cock or into a speckled hen, into a fly or a maize plant, a rhododendron, a beetle, a mouse or a woman. (Schrödinger 1944: 21)

Schrödinger did not have access to the details of how proteins are produced; instead, he gestures towards the hypothesis that the chromosomes contain some kind of code, which in turn is a way of saying that their configuration alone would allow one to predict how development will proceed. Rahul Rose (2012) has pointed out that this type of talk was often explicitly metaphorical, at the same time as it offered plausible interpretations in more neutral biochemical language. So (borrowing Rose's own examples), Goldstein and Plaut hedged their use of semantic language within scare quotes: 'It has been suggested that RNA could serve as a receptor of a "code" from DNA in the nucleus and could transmit this specificity to cytoplasmic proteins, with the synthesis of which it may be associated' (1955: 874). Levinthal stressed the conjectural nature of the code hypothesis when he claimed that 'although there is some evidence that a code exists which translates genetic information into amino-acid sequence, there is no information as to the *nature* of this code' (1959: 254, emphasis in original).

Informational language initially found its way into genetics via a bold and contentious hypothesis, namely that the chromosomal material somehow embodies a code, which specifies amino-acid sequence, or perhaps protein structure. The problem, then, was to find out whether there was indeed such a code, and what that code was. The heritage of the notion of cultural information is quite different. Cultural evolutionary theorists talk of cultural information for three main reasons. First, cultural evolutionists focus in large part on changes in human mental states over time. Mental states are paradigmatic information-bearing states: it is hardly problematic to think that knowledge, for example, involves the possession of information. Second, cultural evolutionary theorists note that one of the things our species is adept at is acquiring knowledge from others through learning processes: so talk of the transmission of information is entirely natural. Third, cultural evolutionists have a significant interest in understanding the processes by which knowledge of various forms is retained and changed: it is not surprising that this becomes framed, informally, as a project to uncover the flows of information that allow the sustenance of cognitive capital (e.g. Sterelny 2012).

Cultural evolutionists spend much of their time constructing models, with the aim of understanding how a collection of intentional states changes under the influence of various forms of learning. That is why, in lieu of offering a definition, Mesoudi instead says, entirely reasonably, that information is 'intended as a broad term to refer to what social scientists and lay people might call knowledge, beliefs, attitudes, norms, preferences, and skills, all of which may be acquired from other individuals via social transmission and consequently shared across social groups' (2011: 3). Much of Richerson and Boyd's work makes a similar compromise: they frequently use the term 'cultural variant' as a shorthand for 'information stored in people's heads' (2005: 63). But they also offer the alternative of 'the ordinary English words idea, skill, belief, attitude, and value' (ibid.). Much of their work seeks to model the ways in which different learning biases can affect the populational profile of a collection of ideas, skills, beliefs, attitudes, and so forth. 'Cultural information', when used in this context, is nothing more than a shorthand for this broad collection of phenomena. This project requires no formal specification of what cultural information is supposed to be and, as we have seen, when theorists are tempted to offer such a specification, it typically lands them in trouble.

3.5 Two Faces of Cultural Information

I have just argued that 'cultural information' is often used as a simple shorthand to denote a variety of intentional states, including beliefs, skills, preferences, and so forth, which are the targets of cultural evolutionary modelling. If this were all that talk of 'cultural information' amounted to, then it would hardly be necessary to devote a whole chapter to its analysis, nor would it be helpful to juxtapose cultural and genetic conceptions of information. Moreover, the informational conception of culture would add nothing to Sperber's superficially contrasting conception of culture as a collection of representational states.

As he puts it: 'Our individual brains are each inhabited by a large number of ideas that determine our behaviour . . . Culture is made up, first and foremost, of such contagious ideas' (Sperber 1996: 1). There is, however, more to the informational conception of culture than this. In Section 3.3 I also suggested that information talk is attractive to evolutionists whenever development, or behaviour, can be interpreted in such a way that it is guided in a suitable manner by the presence or absence of material factors. 'Cultural information' is used in both senses: it is a two-faced term.

Recall that Richerson and Boyd tell us that cultural information need not be stored wholly in people's heads:

> Undoubtedly some cultural information is stored in artifacts. The designs that are used to decorate pots are stored on the pots themselves, and when young potters learn how to make pots they use old pots, not old potters, as models. In the same way the architecture of the church may help store information about the rituals performed within. (2005: 61)

Sometimes, cultural evolutionists use 'cultural information' solely as a label for a phenomenon whose reliable reproduction and adaptation they aim to explain. They want to understand how it is that collections of intentional states are sustained and refined from one generation to the next. At other times, they use 'cultural information' to point to the potentially far broader range of resources that enable this reproduction to occur. Here, 'cultural information' encompasses not only the intentional states of others from whom individuals learn, but also the markings on old pots that enable the skilful laying down of a new set of marks on a new pot.

In this mode, the search for cultural information is the search for those factors that explain the organism's dispositions to behave appropriately. Loosely, it is the search for factors that help to answer Shea's question, 'How does the organism know what to do?' This might involve an appeal to formal teaching, or to the imitation of others, or to the existence of suitably structured learning environments, or to the material constitution of artefacts that invites appropriate use, or to access to written documents, or perhaps to other socially mediated processes we have not yet thought to articulate. 'Cultural information' here acts in an open-ended manner, as a heuristic prompt.

The reliance on a vocabulary of information in this mode implies nothing in particular about whether the mind is best thought of in computational terms, whether a distinction between mental 'hardware' and 'software' is useful, whether developmental outcomes are simple and inevitable results of the 'information' that guides them, whether learning from another involves active engagement, and so forth (cf. Toren 2012). Sterelny's investigations, for example, are framed in terms of information flows, but his project is to understand how and why skills appear and are transformed from one generation to the next, in a manner that often stresses, for example, the creative activities of juveniles, albeit in environments structured by the activities of others. Informational conceptions of the gene have sometimes been attacked on the grounds that they mask essential developmental investigations: by describing genes as bearers of information, we hide the relational activities whereby organisms of various ages bring about their own future developmental stages (cf. Oyama 1985, Thompson 2007: 185, Toren 2012). The informational conception of culture should not be accused of the same fault, precisely because of the heuristic functions of information talk within this domain (Jablonka 2002).

At this point we are forced to raise a series of questions whose answers must wait until Chapter 4. If the search for cultural information is the search for resources that enable the reliable reproduction of various skills and mental states, then a full articulation of these resources will appeal to the distribution of genes as much as it appeals to the distribution of opportunities to learn. It would be tempting, then, to say that 'cultural information' labels the *cultural* resources that enable these forms of reproduction. But how are we to say which resources are the cultural ones? For someone to acquire successfully the ability to milk a cow it is not enough that he can witness a skilled milker: he needs access to a cow, he needs coordination, suitable musculature in the hands, and so forth. The make-up of cows' udders has manifestly changed through the action of conscious and unconscious forms of artificial selection, access to livestock is itself under social control, and coordination and musculature are affected by various forms of training that are in turn subject to social influence. When Richerson and Boyd say that cultural information is partly contained within the structure of a church, they mean that the material layout of the church, itself shaped by builders and users, helps to explain the reliable performance of rituals within it. To turn back to a skill like milking, it now seems we must say that cultural information is contained within cows' udders, within the actions of family and friends who have prompted the acquisition of appropriate motor skills, and so forth. These pronouncements do sound strange, but these awkward consequences are hardly damaging to a broad cultural evolutionary theory. Such a theory seeks to document the ways in which socially structured environments interact with, and reciprocally contribute to the fashioning of, learning dispositions that in turn enable the transmission of valuable skills and beliefs from one generation to the next.

4

Human Nature in Theory

4.1 Human Nature, Human Culture

Many philosophers of biology today think about human nature in the same way that they think about race. Race is commonly regarded as having no basis in biological reality, and the same, it is sometimes said, goes for human nature. This sort of sceptical stance about human nature is perhaps most strongly associated with seminal work by David Hull (1986). And yet, there are plenty of workers in the biological and human sciences whose remarks seem to presuppose that human nature is a real object of investigation, and that they are well placed to describe and understand it. This is most evident when we look to popular science. Steven Pinker's (2002) book *The Blank Slate: The Modern Denial of Human Nature* intimates by its title that that 'human nature' should not be denied. Prominent remarks by pioneering evolutionary psychologists reinforce this image of a set of sciences that aim to say what human nature is. In 1990, Leda Cosmides and John Tooby announced their intention to defend 'the concept of a universal human nature, based on a species-typical collection of complex psychological adaptations' (Tooby and Cosmides 1990: 17). Meanwhile, observers from historical and social anthropological schools of thought have felt that these images of human nature are deficient. They have sometimes argued that evolutionary accounts of human minds and human cultures are imperilled precisely because of their commitments to such views of human nature. We have already encountered Roger Smith's worry that evolutionary approaches 'continue to reflect the belief that there is an essential human nature' (Smith 2007: 27).

It might seem odd that a book about cultural evolution includes two chapters devoted to the topic of human nature. After all, it is primarily evolutionary psychologists of Cosmides and Tooby's school, rather than cultural evolutionists, who have made overt use of the human nature concept. Peter Richerson, probably one of the most influential evolutionary theorists of culture, has sometimes written to fellow evolutionary thinkers recommending that they simply stop using the concept of human nature altogether.

Let me be clear, then, that I will not be arguing that cultural evolutionary theory has a deep commitment to any strong notion of human nature; quite the opposite. Recent work in cultural evolutionary theory offers us some of the best reasons for opposing efforts to articulate substantive notions of human nature. But it is worth discussing human nature in some detail, in large part because some of the main suspicions of the cultural evolutionary approach derive from a perception that it is somehow committed to a strong distinction between human nature and human culture. Maurice Bloch (2012: 1), for example, seems to have the impression that struggles regarding the legitimacy of applying evolutionary thinking to human minds and human culture are struggles over the very idea of human nature. In this chapter we will see several flaws in recent efforts to articulate substantive accounts of what human nature is. But we will also see that cultural evolutionary theorizing requires no such notion of human nature, and so it is not imperilled by this scepticism.

This chapter begins with a brief overview of Hull's philosophical argument against the very notion of human nature. I then move on to criticize three recent philosophical attempts to salvage the scientific respectability of human nature, which we owe to Edouard Machery, Richard Samuels, and Grant Ramsey. My criticisms draw attention to two phenomena that pose problems for all but the most permissive accounts of human nature. First, evolutionary explanation focuses just as much on variation within a species as it does on uniformity. The very same processes—selection, mutation, drift, and so forth—that can result in traits being present at near to 100 per cent can also result in a variety of stable polymorphisms. So there is no good pragmatic reason to restrict the attention of evolutionary scientists to those traits that are universal. Second, recent work by the likes of Cecilia Heyes and collaborators has established the ways in which various forms of learning play important roles in the development of cognitive mechanisms—especially mechanisms of imitation—that are of central importance for the historical trajectory of our species. So there is also no good pragmatic reason for evolutionary scientists to restrict their attention to traits whose development is largely independent of the processes of social learning.

The relevance of our discussion of human nature to the defensibility of cultural evolutionary work will become clearer still in Chapter 5, and so a short preview of that discussion is in order. How are we to understand the relationship between nature and culture? Some of the most respected work in cultural evolution articulates the explanatory and causal roles of cultural forces through models of *gene/culture co-evolution*. Cultural changes, so these stories go, can have knock-on effects on genetic evolution, as when the spread of dairying through *cultural* transmission encourages the subsequent spread of *genetic* mutations, under the action of natural selection, which promote lactose tolerance.

Maurice Bloch, among several others, objects to these models of gene/culture co-evolution on the grounds that culture and genetics are not distinct forces that can influence each other, but instead need to be thought of as 'a unified process' (2012: 52). In other words, critics hold that cultural evolutionists endorse a problematic distinction between nature and culture, manifested in their claims about how natural and cultural processes interact. Criticisms of this sort have been common among social anthropologists, but they are also present in the work of philosophers, biologists, and psychologists sympathetic to developmental systems theory (DST), an approach that will be explained in more detail later in Chapter 5. The semblance of a nature/culture distinction in important cultural evolutionary work also lends plausibility to the idea, endorsed by Machery and Samuels, that the behavioural sciences presuppose that some traits are indeed elements of human nature, while others are elements of human culture.

In Chapter 5 I will argue that sceptics are right to reject very strong distinctions between the natural and the cultural. I will also argue that this scepticism leaves the cultural evolutionary project intact. A close examination of models of gene/culture co-evolution, and dual-inheritance theories, shows that in spite of their names they are not committed to strong views about the independence of genetic and cultural causation. Instead, they are compatible with integrated images of organic development and evolution, which reject alleged oppositions between the innate and the acquired; between that which is genetic and a product of natural selection, and that which is social and a product of cultural transmission; between cultural and genetic channels of inheritance.

4.2 A Recent History of 'Human Nature'

David Hull's (1986) paper gives the classic philosophical case against human nature. He was sceptical of the very notion, in large part because he assumed that human nature, if there were such a thing, would need to be analogous to the alleged essences of chemical elements. On one very influential story, the possession of seventy-nine protons is both what makes something a sample of gold (rather than some other element), and also what explains why samples of gold have characteristic properties. Hull assumed that 'human nature' would also have to do two jobs at once. First, it would have to pick out some underlying aspect of the constitution of individual organisms that makes them humans (rather than cats, or dogs, or potatoes). Second, it would also have to explain why these organisms behave in characteristically human ways. Modern biology, he argued, denies that any property, or set of properties, plays both of these roles, and it thereby denies that human nature is real. These sceptical arguments against human nature are well known and it would be tedious to rehearse them in any great detail, but it is worth offering a precis of some important moves in the debate.

In large part, Hull's scepticism of human nature derives from his views about taxonomy. Hull took it that species are to be defined phylogenetically. I will not question this assumption here, although alternative views of taxonomy are more popular in some biological circles than one might think (see Lewens 2012a). Hull says that a species is a complex individual in its own right, and more specifically, it is a twig on the tree of life. In asking what makes an organism a member of a given species, we are asking a question about what determines part-hood in a larger individual. In general, these matters are decided by appeal not to microstructural commonalities, but to relational orientation. The parts of a lamp can be composed of very different materials: what makes them parts of a single larger object is a matter of how they are stuck together, not a matter of their being made of similar sorts of stuff. Likewise, according to Hull, organisms are parts of a species not by virtue of their common intrinsic properties, but instead by virtue of the genealogical relations they stand in to each other. Hull's manner of approaching the question of human nature links it intimately to questions about essentialism and taxonomy: to deny (as many have done) that species membership is determined by the possession of a characteristic set of intrinsic properties simply is (for Hull) to deny that species have natures, and a fortiori it is to deny that there is such a thing as human nature.

In recent years there have been two developments that challenge Hull's position regarding human nature. First, Michael Devitt (2008, 2010) has tried to argue that species do have intrinsic essences after all. I have discussed Devitt's views at length elsewhere (Lewens 2012b), and I will not devote time to them here. Second, a variety of recent commentators have observed that we should separate the question of what human nature might be from the question of what determines membership of the human species. The great majority of humans have two legs; to assert this is not to assert that someone with only one leg is not human. It is plain that one can consistently deny that species membership is determined by the possession of some set of intrinsic properties, while also asserting that there are many interesting generalizations to be had about what sorts of properties are typical of a given species, and about what sorts of processes explain why it should be the case that such properties are typical.

Even if we agree that members of our lineage might change in any respect, and to any degree, while *Homo sapiens* continues to exist, and even if we agree that no trait is, at the current moment, present in all and only humans, we might still think that natural selection has made it the case that certain developmental processes, or psychological dispositions, are present over reasonably long periods in *almost* all humans. Hull himself rather grudgingly acknowledged that such a conception of human nature might be available, but he seemed dismissive of it: 'If by "human nature" all one means is a trait which happens to be prevalent and important for the moment, then human nature surely exists' (1986: 9). The job for the remainder of this chapter is to examine three proposals for a more biologically respectable account of human nature, all of which deny that the reality of human nature is settled by our views about taxonomic essentialism. All three views seem reasonable on first inspection, but I will claim that every one of them encounters significant problems. Hull was right to be dismissive of these more deflationary pictures of human nature.

4.3 Machery's Nomological Conception

Edouard Machery (2008, 2012) has argued for a biologically legitimate notion of human nature, the detailed description of which he takes to be the goal of work in evolutionary psychology. He follows Hull in dismissing the thought that the possession of human nature determines whether some organism is a member of *Homo sapiens*. In place of this essentialist view he instead offers what he calls a 'nomological conception', which has it that 'human nature is the set of properties that humans tend to possess as a result of the evolution of their species' (2008: 323).

Richard Samuels (whose own very closely related account we will examine in a moment) reckons that Machery's account is 'relatively undemanding' (2012: 13), but that is because Samuels downplays the problems associated with two key elements of the position. It is indeed undemanding, metaphysically speaking, to suggest that there might be properties that many humans possess. Machery's account does not stop there. In addition to requiring that something is only a part of human nature when it is 'shared by most humans', he adds that something is only a part of human nature when humans possess that feature 'as a result of their evolution' (2008: 323). So Machery's account faces two rather demanding tasks: the first is to explain why it should be theoretically important for some group of sciences to focus on those traits that are present at high frequencies, rather than also taking an interest in traits that are present at lower frequencies. The second is to say how we are to demarcate the processes that are 'evolutionary' from those that are not. I contend that neither job can be done in a way that is favourable to Machery's cause.

Machery takes it that 'human nature' picks out a suitable object of study for a subset of the sciences. He suggests that it is important that various types of psychological and behavioural sciences, for example, should focus on documenting and accounting for those traits that are present in most humans. But what makes this a special explanandum? Consider that for the evolutionist, natural selection, mutation, and drift can result in traits that are present in populations at near 100 per cent ('fixation'), but they can also bring about a range of stable polymorphisms. From the evolutionary perspective—as David Buller (2005) has stressed in his sceptical treatment of evolutionary psychology—there is no special significance attached to traits that are at, or near to, fixation, and the explanatory processes that evolutionists deal in are just as suited to accounting for traits that are present in different proportions.

Not only are these explanatory processes potentially able to account for traits present at any frequency in a population, but the enquiry into the maintenance of polymorphism is central to the work of various branches of evolutionary enquiry. Examples abound: evolutionary game theory typically examines the circumstances under which numerous alternative strategies may coexist in a population; evolutionary psychology sometimes hazards distinctive psychological profiles for males and females; more direct empirical enquiry documents the fascinating array of distinct morphological and behavioural strategies that may coexist within a given species.

A fairly standard textbook example concerns the side-blotched lizard (*Uta stansburiana*). Three alternative morphological and behavioural forms coexist in the males of these populations. Orange-throated males are very aggressive, and defend large territories; dark blue-throated males defend smaller territories, and are correspondingly less aggressive; males with yellow-striped throats do not defend territories at all, and instead adopt a 'sneaky' mating strategy. Sinervo and Lively (1996) argue that the fitness ordering of the three strategies follows a 'scissor-paper-stone' pattern, which means that no strategy ever goes to fixation. Orange throats are beaten by yellow stripes, yellow stripes are beaten by blue throats, and blue throats are beaten by orange throats. Hence, while the frequencies of these different strategies may oscillate, they are all maintained at significant levels.

The marine crustacean species *Paracerceis sculpta* constitutes a very similar example (Shuster 1987). Here again there are three male forms: one large one that guards smaller females in harems contained inside sponges, one very small one that sneaks into the sponges, and one that imitates the females to enter the sponges in disguise. Again, selection maintains all three.

Machery's first basic problem, then, is to explain why, if 'human nature' is supposed to fix the subject matter of (among other things) evolutionary psychology, the term should only name traits that are universal in our species. He defends this decision on purely pragmatic grounds: since other disciplines (he names forms of anthropology, and personality psychology) focus on human differences, 'it is useful to have a notion that picks out the similarities between humans' (2008: 324). True enough, personality psychology and social anthropology document diversity. But they are rarely in the business of offering evolutionary explanations, and a science that does offer evolutionary explanations will be impoverished if it focuses only on traits that are present at fixation.

If Machery's aim in outlining a respectable concept of human nature is to ensure that an important realm of inquiry is not neglected, then the study of evolved polymorphisms would seem just as worthy of support as the study of traits that evolution has distributed more widely in our species.

Machery has recently responded to this challenge in a slightly different way. He suggests that 'if one wants a useful notion of human nature, upholding the universality assumption may be the only way to avoid drawing an arbitrary line' (2012: 477). We have seen, though, that evolutionary processes can result in traits that go to fixation, and they can also result in traits that appear at any other frequency. So, from an evolutionary perspective, a focus on traits that are present at near to 100 per cent is really no less arbitrary than a focus on those traits that are present at 30 per cent or 90 per cent. Machery also makes no case here for what sort of job a 'useful notion of human nature' is supposed to do: it begs the question to say that what makes a notion of human nature useful is that it picks out those traits that are near universal. As we have seen, if a human nature concept discourages attention to evolved polymorphism, then its practical value is diminished.

This brings us to the second problem for Machery's account. Not just any common trait is part of human nature on his view, only those present as 'a result of evolution'. Which traits are these? The problem is compounded by his own insistence that:

nothing is said about the nature of the evolutionary processes in the proposed characterisation of human nature. The traits that are part of human nature can be adaptations, by-products of adaptations, outcomes of developmental constraints, or neutral traits that have come to fixation by drift. (2008: 324)

Of course, this does not go very far to explaining what does count as an evolutionary process. Machery understands that at least some constraints must be placed on this feature of his account, and he consequently proposes that: 'saying that a given property ... belongs to human nature ... is to reject any explanation to the effect that its occurrence is exclusively due to enculturation or to social learning' (2008: 326).

There are a number of ways in which we can put pressure on this suggestion. First, Machery's thought appears to be that social learning is not itself an evolutionary process. Hence, if social learning is the only process responsible for the occurrence of some trait, the trait does not count as a part of human nature. To make this argument good one needs some reason for denying that evolutionary processes in general include cultural evolutionary processes. As we have already seen, the likes of Richerson and Boyd (2005) have built fruitful evolutionary models that consider how populations change over time—how they evolve, that is— under the influence of various forms of social learning. It is plain that social learning can be a cumulative constructive force: it is because of our ability to learn from others that over time we have developed complex techniques, technologies, and theoretical stances that profoundly affect how we think and how we live. So, social learning cannot be written off as a mere background change in the environment that affects how genetic adaptations are expressed. Second, if Machery's underlying intuition is that 'human nature' should name those traits that are widely distributed among our species, and whose development is very robust, or hard to evade, it is not clear why we should rule out an explanation for their persistence that looks solely to enculturation or social learning. The question of whether widely distributed traits, which are hard to evade, might develop as a result of learning should be one that is open to empirical enquiry, rather than one closed off by definitional fiat.

Third, it is not clear that Machery's conditions succeed in excluding any widely distributed traits from human nature. He gives an example of a trait that he thinks is not part of human nature: 'the belief that water is wet is not part of human nature, in spite of being common, because this belief is not the result of some evolutionary process. Rather, people learn that water is wet' (2008: 327). Why should we not count this belief as the result of an evolutionary process? Machery suggests two rather different answers, but neither is satisfactory. The first response looks back to the notion that if a trait's occurrence 'is exclusively due to enculturation or to social learning', then it is not evolved. We have already seen reasons to question this stipulation, but Machery is quick to point out that 'This is of course not to deny that social learning, or indeed any other environ-mental influence, can be part of the explanation of the development of the trait' (2008: 328, fn. 8). Once we concede that a ubiquitous trait can be part of human nature so long as social learning does not provide a full explanation for its occurrence, then we will judge that since part of the development of the belief that water is wet refers to phenomena other than enculturation and social learning, such as one's own perception of water's wetness, this ubiquitous belief can qualify as part of human nature after all.

Machery's second response is rather different:

Saying that a trait has an evolutionary history is to say something stronger than the fact that it has perdured across generations. Humans have probably believed that water is wet for a very long time, although this belief has no evolutionary history. For this trait is not a modification of a distinct, more ancient trait. (2008: 327)

This may be a plausible claim for beliefs about water, but it looks less plausible if we consider a different example of a trait that is very widely distributed among the human species today, such as basic knowledge of the rules of association football ('soccer'). This knowledge has surely undergone modification over time, as the rules of the game have changed. If we follow Machery's second response, knowledge of the rules of association football is part of human nature, while knowledge of the wetness of water is not. Machery's proposal to equate human nature only with those elements of our common make-up that can be understood as modifications of earlier, more ancient traits threatens to draw a theoretically dubious distinction between different cognitive traits, all of which are produced via social learning.

So far, the objections I have presented are largely of an a priori nature. Since it might turn out that entrenched forms of learning, which are widely distributed around human societies, could give rise to more or less universal traits that are also very difficult to get rid of, there is no especially good reason to use 'human nature' to name traits whose development does not rely primarily on social learning. Because this is an intriguing empirical possibility, cognitive and evolutionary sciences should be open to examining it.

Machery might respond in a rather impatient manner: it is reasonable to offer a restrictive definition of human nature, so long as that definition has empirical support. Unless we have good experimental or theoretical reasons to think that learning might actually be responsible for some widely distributed and reliably appearing traits, we are justified in keeping a more restrictive definition of human nature.

At this point we can profitably turn to empirical and theoretical work by Cecilia Heyes and collaborators. Human social learning is usually understood as the ability to learn from observing other humans, through channels other than explicit teaching. Heyes argues that this ability makes use of general associative learning capacities (the sort of capacities that underpin individual trial-and-error learning, for example), coupled to cognitive input mechanisms specifically tuned to the activities of other agents (Heyes 2012a, 2012b). Importantly for our purposes, Heyes suggests that this tuning of input mechanisms—that is, the shaping of perceptual, attentional, and motivational systems that direct us to the activities of other people—may itself be influenced by learning (see Ray and Heyes 2011, Heyes 2012a). Among other things, she suggests that young children learn to attend to specific elements of other people's faces, for example, when assessing emotional states of others. She also suggests that imitation—a specific form of social learning that involves copying the actions of others—may be a learned capacity.

One puzzle for theories of imitation is how the 'correspondence problem' is solved. An imitator needs to witness an action in another, and then produce a similar one. The problem is that how an action looks to an observer does not resemble how it feels for someone executing the action. The puzzle is especially acute when one's own bodily movements are hard to observe. It is not at all clear how a child who sees someone else move his mouth in a certain way is able to copy that action, because she cannot see her own mouth to check on how it is moving, and the feel of her own mouth moving does not resemble the look of another's mouth moving in the same way.

Heyes's associative sequence learning (ASL) model proposes that the links between perception of an action, and execution of the same action, can indeed be learned, so long as the two are reasonably reliably associated. They can come to be associated in part because individuals can look at their own actions, in part through the use of artificial supports such as mirrors, in part through witnessing action patterns in others when one is part of a group engaged in common tasks, and so forth. Heyes argues that these perception/execution correlations bring about learned linkages between sensory and motor representations of simple units of movement. Once these links are established—i.e. once the correspondence problem has been solved via learning—then further imitation is possible when more complex patterns of movement units are observed together.

Is there not a widespread view these days that imitation is achieved through the action of 'mirror neurons'? Heyes does not deny the existence of mirror neurons; instead she argues that their mirroring properties are explained as an outcome of associative learning, as spelled out in her ASL model (Heyes 2010). Of course, this view does not tell us that learning is the process we must appeal to in the development of every aspect of the ability to imitate. One might agree with Heyes, for example, that infants learn linkages between sensory and motor representations of hand movements, while adding that their ability to learn is facilitated by an innate tendency to look at their own hands. So Heyes need not deny the propriety of all appeals to 'innate' capacities in the explanation of imitation. But she does suggest that when we ask the question 'How is it that babies acquire the ability to perform an action, after witnessing another performing the same action?', the answer must appeal to learned associations between sensory and motor representations. It is in this sense that she argues that imitation is a learned capacity.

I do not mean to suggest that Heyes's view is secure, but it does have significant empirical backing. It helps to explain the facts that chimpanzees can be trained to imitate, that imitation in newborn human infants appears to be restricted to tongue protrusion at best, that birds seem able to imitate behaviours that they engage in collectively as flocks, and so forth (Heyes 2001, 2010). This all means that we should take seriously the thought that learning might be responsible for the ontogeny of a trait—in this case the capacity to imitate others—that is at one and the same time widely distributed, hard to eliminate, and of great significance for human interactions. Imitation is just the sort of trait that one would presumably want to count as part of human nature, and the question of the role of social learning or enculturation in its production is one that needs to be explored in full.

4.4 Samuels's Essentialism

I will devote only a few paragraphs to Richard Samuels's recent proposal for how to understand 'human nature', for his account is very close to Machery's, and it inherits many of its problems (Samuels 2012). Samuels shares with Machery a conviction that the description of 'human nature' is an important goal for a set of sciences, including cognitive science. He suggests that, if possible, we should strengthen Machery's account to enable it to discharge an important desideratum of any account of human nature. On the face of things, says Samuels, natures are explanatory. Machery's account, by contrast, allows that mere patterns of behaviour (so long as they have the right sort of evolutionary history) will count as elements of human nature.

Samuels's own 'causal essentialist' account aims to rectify this problem by requiring, like Machery, that 'human nature' should name statistically common features within the human species, but by adding further conditions to the sorts of features that qualify. Samuels requires that 'human nature should be identified with a suite of mechanisms, processes, and structures that causally explain many of the more superficial properties and regularities reliably associated with humanity' (2012: 20).

Samuels's use of the label 'causal essentialist' to qualify his account may mislead some readers. He is not committed to any strong form of modal essentialism here. He is not claiming, for example, that no organism could possibly be a human unless it instantiated the relevant underlying causal mechanisms. His view explicitly allows that some small number of actual humans in fact lack the causal mechanisms that are constitutive of human nature. All Samuels means by 'essences' are 'entities—mechanisms, processes and structures—that cause many of the more superficial properties and regularities reliably associated with the kind' (ibid.).

Samuels's view is close to that of Machery because he requires that 'human nature' picks out statistical commonalities at the level of cognitive mechanisms, rather than at the level of the equally common behaviours that such mechanisms might produce. Once we have a catalogue of widely found cognitive 'mechanisms, processes and structures'— anything from frequently encountered patterns of neural development through to statistically typical heuristics for estimating probabilities— then we have discharged the task of discovering the 'essences' that lie behind the regularities in our behaviour. The upshot of this is that Samuels's account, just like Machery's, faces worries about why we should privilege those cognitive, neural, or developmental mechanisms that are common over those that are present at lower frequencies in the population, when we ask what is comprised by 'human nature'.

I have assumed that Samuels simply adds a further causal-explanatory condition to Machery's own conception of human nature, but some of Samuels's remarks undermine that interpretation. First, Samuels does not mention Machery's requirement that human nature names those traits that are 'a result of evolution'. Instead, he focuses solely on the requirement of statistical commonality, suggesting that 'a theory of human nature should capture aspects of human beings that are in some sense species-typical' (2012: 10). Samuels's own adherence to this condition is, however, doubtful. In a footnote, he mentions the importance of regularities of difference: our species predictably produces variation, as well as predictably producing similarity. He lists sexual dimorphism and 'systematic behavioural or morphological variation'. Confusingly, he notes that he sees 'no serious problem with accommodating such regularities into an account of human nature' (2012: 10). But once we dispense with a requirement that the focal regularities for an account of human nature need to be universal, and once we embrace neutrality regarding the sorts of processes that explain their regular appearance, then our conception of human nature becomes wholly unconstrained. More or less any causal-explanatory mechanism, regardless of how local it might be, will count as constitutive of human nature. Many, but certainly not all, humans are able to read. Some set of characteristic neural mechanisms, which are generated through learning processes, explains this widely distributed ability (Heyes 2012a). These mechanisms, according to Samuels, must then count as elements of human nature. Once we agree to include specialized mechanisms for reading as a part of human nature, it is hard to see why we would not exclude specialized neural mechanisms that underpin driving, or any other acquired skills.

4.5 The Field Guide Error

Both Machery and Samuels have been attracted to an analogy between descriptions of human nature and the sort of description one encounters in field guides. Machery has subsequently repudiated this link (2012), but I will say a little about it here, because one might be under the impression that since the production of field guides is manifestly a respectable activity, then something like Machery's or Samuels's conception of human nature must be scientifically respectable, too. As Machery formerly put things, 'describing human nature is thus equivalent to what ornithologists do when they characterize the typical properties of birds in bird fieldguides' (2008: 323). Samuels, too, remarks that 'We should think of a theory of human nature as, amongst other things, providing a kind of field guide to humanity' (2012: 9). But the field guide analogy is misleading.

An ornithological field guide is designed to enable the correct recognition of an individual bird, qua member of a given species or subspecies, in the field. Since this is the function of field guides, it is not surprising that they restrict themselves to properties that are (a) diagnostic of a given species, and (b) easy to observe over a short period of time (such as when out on a walk). This is one of the reasons why ornithological field guides often aim to describe birds' calls. A call can be diagnostic of the bird's species even when it is learned. Here, then, is a property that does feature in a field guide, but does not fall under Machery's concept of a species' nature.

There are also many properties that do not feature in field guides, but which do fall under Machery's concept of a species' nature. Recall that on Machery's nomological conception, a property can be part of human nature when it is neither diagnostic of our species, nor easy to observe: the property instead needs to be widely distributed through our species. Some such properties may also be found widely among other species (hence, they are not diagnostic), and they may only be apparent in an individual human after microscopic observation, or prolonged observation (hence, they would not be easy to observe).

Machery's conception of human nature is supposed to make sense of, among other things, the explanatory targets of evolutionary psychology. Suppose, then, that evolutionary psychologists Daly and Wilson (1988) are correct in their view that humans treat genetically unrelated offspring unfavourably. This would be part of 'human nature' on Machery's view in spite of the fact that it is not the sort of feature one can easily observe on a short walk in the field, and it is not the sort of feature that is peculiar to humans: on both counts, it is not the sort of feature one would expect to find in the entry under 'Human' in a field guide to mammals. Much more generally, the vast number of homologies shared between humans and related species (pentadactyly, the possession of a spinal cord, and so forth) are parts of human nature on Machery's view, even though they are not diagnostic of our species.

Finally, it is worth remembering that field guides do not pretend to be technical documents summarizing the full range of variation that might occur among species members, and which would thereby allow thoroughly reliable diagnosis. The images and descriptions in these guides instead serve as heuristics for use either by amateurs, or as a precursor to a more technical application of a species key. It is misleading to say that the description of human nature under the nomological conception is equivalent to the sort of description one finds in an ornithological field guide.

4.6 Ramsey's Patterns

As a final example of a recent effort to make human nature scientifically respectable, I consider the 'life-history trait cluster' (LTC) account of Grant Ramsey (2013a). Ramsey's account has two clear virtues when compared with the accounts of Machery and Samuels. First, it acknowledges that variation is no less significant than uniformity. Second, it deliberately eschews any problematic appeal to a distinction between those traits that are innate and those that are acquired. In spite of all this, we can ask whether Ramsey's account is as disciplined as he would have us believe.

Ramsey's critique of Machery is consonant with my own. A respectable account of human nature should not focus on what is typical, nor should it be committed to the dubious view that enculturation will play no role in the ontogeny of developmentally robust traits. Ramsey notes instead that, if we survey the entire human population, we will see various 'patterns' within it. Some traits will be common: almost all humans are bipedal. Some will be rare: hardly any humans have mastered tensor calculus. Ramsey himself is especially concerned with the sorts of conditional patterns that can be expressed using what we might call 'Ramsey conditionals': 'If someone grows up in the UK, then they tend to learn English;' 'If someone eats a poor diet, then they are likely to encounter health problems.' The first of these conditionals Ramsey calls 'robust', since the association between antecedent and consequent is fairly reliable. It is not 'pervasive', since (on a global scale) the antecedent itself is rarely satisfied. The second conditional is both robust and far more pervasive, since the association it picks out is fairly reliable, and its antecedent is often satisfied.

The problem with Ramsey's view is that it is, in his own words, 'extremely inclusive' (2013a: 987). Perhaps he sees this as a strength, rather than a weakness: we agree that the best account of human nature that can be constructed will also be exceptionally capacious. To see just how capacious Ramsey's account is, consider that he tells us simply that 'Human nature is defined as the pattern of trait clusters within the totality of extant human possible life histories' (ibid.). At this point Ramsey gives no constraints on what is meant by 'pattern': since it is a reliable feature of humans that hardly anyone masters tensor calculus, one might wonder if the rarity of mastery of tensor calculus is itself part of the 'pattern' of extant human possible life histories.

Ramsey does not require that bona fide elements of human nature be expressed by conditionals that are both robust and pervasive. Instead, because he rightly notes that many robust and pervasive conditionals will be vacuous ('If one is born, then one has mass'), he argues that we should aim for conditionals that are informative. Many informative conditionals will sacrifice pervasiveness for robustness:

> instead of saying that 'lactation is a part of human nature,' one could say that 'lactation is a part of female human nature' or that 'female lactation is a part of human nature' since the latter descriptions pick out an antecedent (being female) that makes the antecedent-consequent association robust. (2013a: 991)

Ramsey adds that he 'will define uniquely human nature as the subset of the antecedent-consequent associations that are unique to the human species' (2013a: 992).

When we put these claims together, then we see that a sentence such as 'If one grows up in New Zealand, then one learns the rules of rugby union' identifies an element of uniquely human nature, for it picks out a robust association that is unique to the human species. The non-human animals of New Zealand do not learn the rules of rugby union. Likewise, a conditional such as 'If one grows up in China, then one learns to eat with chopsticks' is fairly robust, it is fairly pervasive (given the large Chinese population), and it is uniquely human (since no other animal uses chopsticks to eat). The same holds true for the conditional 'If one grows up with access to a television, then one can recognise David Beckham.' For Ramsey, these last two conditionals will both score highly as informative claims about human nature.

Ramsey does not seem especially troubled by these aspects of his account. In response to critics he points out that a claim such as 'If one is raised in America, then one drives on the right' tells us something informative, and something robust. It also tells us something that sets humans apart from other species (Ramsey 2012: 482). There have been reports of dogs learning to drive in New Zealand (where the norm is to drive on the left), but to my knowledge no non-human animal raised in America drives at all, and a fortiori not on the right. For Ramsey, a description of human nature is simply a collection of informative truths about humans. His account demonstrates that extreme liberality is the price of defensibility in this domain.

4.7 Our Libertine Natures

What lessons can we take away from our discussion of these proposals for respectable accounts of human nature? At first sight they seem reasonable. Of course there are fairly reliable generalizations about the cognitive and behavioural traits of our species; of course the distribution of such traits often owes itself, in some way or another, to evolutionary processes. But as we inspect these conceptions in more detail, they slip through our fingers.

There is no particularly good theoretical or pragmatic reason to insist that the 'nature' of a species consists solely in those traits that are widely distributed throughout it. This is especially clear when we focus on the conception of a biological species as an individual; that is, as a complex structured entity in its own right. The human species, like all species, has various reliable tendencies, in virtue of the distribution of genetic and many other developmental resources throughout it (Lewens 2009b). It is in the nature of the human species reliably to produce organisms that are similar in certain respects, but it is also in the nature of the human species as a whole reliably to produce patterns of difference, such as the anatomical and physiological differences between the sexes (Ereshefsky and Matthen 2005). There is also no particularly good theoretical reason to insist that traits cannot count as parts of the species' nature if they are solely produced by forms of enculturation or social learning; that is partly because it is unclear that any traits are produced solely through these processes, partly because learning processes can underpin the development of traits that are reliably distributed throughout the species, and partly because various forms of social learning appear to have been important in the evolutionary history of our species. These comments resonate with Paul Griffiths's 'reconstruction' of the human nature concept: 'human nature', he says, 'is primarily the pattern of similarity and difference amongst human beings. This pattern results from the operation of the evolved human developmental system in a wide variety of environments, some of them novel' (2009: 53).

If these concessions are made, conceptions like those of Machery and Samuels become far more permissive than their proponents intend. If we understand 'human nature' simply to name the reliable dispositions of the human species as a whole, then what grounds do we have to deny that it is in the nature of the human species to produce Sikhs, or that it is in the nature of the human species to produce skiers? These are certainly reliable tendencies of our species, which have endured over a reasonable length of time. We cannot exclude them on the grounds that there have not always been Sikhs, or skiers, because any biologically respectable notion of human nature must allow that our species' nature can change. We cannot exclude them on the grounds that not everyone is a Sikh, or a skier, because many of the most important dispositions of our species involve its ability to produce predictable forms of variation. Nor can we exclude them on the grounds that Sikhism, or skiing, is learned, because any biologically respectable notion of human nature must allow that learning contributes to our make-up. Ramsey understands these problems, and his resultant conception of human nature is exceptionally liberal as a result.

Once an account of human nature is loosened so as to make room for variation and learning, there is no way to regain control of it. Machery takes it that this instability grounds a pragmatic argument for disciplining 'human nature', even if this discipline involves a form of arbitrary fiat. This pragmatic proposal has a number of weaknesses. We are faced with three choices: first, we can simply accept the libertine account of 'human nature', as a name for all the reliable dispositions of our species, from the global to the local. Second, we can argue that we are best off getting rid of the human nature concept altogether. Third, we can argue that we need a disciplined human nature concept, along the lines Machery proposes. I do not propose to adjudicate between the first and second options here. The question of whether human nature is best eliminated, or best understood in an exceptionally liberal manner, will depend in large part on issues about the extent to which the libertine conception I have sketched can plausibly be thought of as discharging the core intuitions that have been historically associated with 'human nature', and also on issues about the harm we might expect to come from misunderstandings of the significance of this libertine notion, if it carries the 'human nature' label (Lewens 2012c). The drawback of the third proposal is that Machery goes only so far as to claim that *if* a disciplined human nature concept is required, *then* his own proposal fits the bill well. But why do we need such a disciplined human nature concept?

It is far from clear that any form of scientific enquiry would be impoverished if we were either to endorse a thoroughly libertine conception of human nature, or if we were to get rid of the concept entirely.

In both cases scientists can still document and explain the dispositions of species to produce traits that are more or less ubiquitous among their members. The libertine conception enables this sort of enquiry in a manner that blinkers us neither to the possibility that ubiquity might be a product of learning, nor to the important ways in which a species can also reliably produce diversity. The libertine concept is the best we can do. In Chapter 5, I argue that it is also good enough for cultural evolutionary investigation.

5

Human Nature in Practice

5.1 A Pragmatic Turn in the Human Nature Debate

We cannot use evolutionary research to argue that 'human nature' should be restricted to those traits that are ubiquitous in our species, nor can we use this research to argue that 'human nature' should be restricted to those traits whose ontogeny does not depend in important ways on learning. We established those claims in Chapter 4. This chapter moves on to look in more pragmatic ways at how cultural evolutionists understand the relationship between human nature and human culture, and more specifically at whether their practices of research and modelling commit them to a problematic conception of that relationship. I argue that, in spite of superficial appearances to the contrary, their handling of these notions is for the most part without significant fault.

This pragmatic turn is already explicit in Machery's recent response to critics, where he suggests that the human behavioural sciences require a constrained notion of human nature, and that they also articulate the distinctions that enable those constraints to be put in place (Machery 2012). Cultural evolutionists do much of their work by fashioning and examining models of gene/culture co-evolution. They tell us, for example, that lactose tolerance is a genetically controlled adaptation, which was favoured by natural selection only after a novel cultural environment of pastoralism became established through social learning (Holden and Mace 1997). Machery takes it that the distinction between what he calls 'organic' and cultural selection, which appears to be presupposed by these co-evolutionary models, gives us the blocks we need to build a wall between those traits that are elements of human nature, and those traits that are elements of human culture.

A ramshackle alliance of theorists from psychology, social anthropology, and philosophy has expressed scepticism with any attempt to distinguish the natural from the cultural. They often draw on developmental systems theory (DST), a theoretical orientation to the biological world first laid out in works by Susan Oyama (1985), Paul Griffiths, and Russell Gray (Gray 1992, Griffiths and Gray 1994), which stresses the great variety of resources that causally combine to produce offspring that resemble their parents, and which is consequently suspicious of talk of distinct genetic and cultural 'channels' of inheritance. Anyone sympathetic to DST might fear that if cultural evolutionists are indeed committed to a nature/culture distinction, then cultural evolutionists are in trouble. Christina Toren (2012: 27) notes that the 'culture–biology' distinction has long been considered problematic, especially by social anthropologists concerned with psychology. Tim Ingold (1995) and Maurice Bloch (2012) both detect troubling appeals to such a distinction in the cultural evolutionary project, and they have dismissed that project as a result.

In Chapter 4 we established that evolutionary considerations license only the most libertine conception of human nature. If the practice of cultural evolutionary theorizing mandates a more restrictive notion of human nature, then we face a troubling dilemma. Perhaps Machery and Samuels are right after all: there is a perfectly respectable, albeit restrictive, concept of human nature, which contrasts with human culture. Or perhaps cultural evolutionary theory is theoretically compromised at its heart, as Ingold and Bloch have suggested. We can reject both of these options. Ingold, Bloch, and the developmental systems theorists are right to oppose naïve efforts to pull nature and culture apart, but none of these worries seriously undermines the practice of gene/culture co-evolutionary modelling. These models have no commitment to strong versions of the nature/culture distinction.

5.2 Who Needs Human Nature?

Does any science presuppose a restrictive notion of human nature, of the sort defended by Machery? Richard Samuels (2012) has recently appealed to some of Noam Chomsky's comments to make a case for thinking that 'human nature' has an important organizing role in the sciences. It is helpful to set Chomsky's remarks in context. They were made during his 1971 debate—aired on Dutch television—with Michel

Foucault. He and Foucault were engaged in a discussion over the very issue of what scientific weight we can assign to a term like 'human nature'.

Foucault had expressed doubts about how the term functioned. He suggested it did not act as a precise technical term, picking out an entity whose constitution is to be probed and elaborated by experiment. One might think a term like 'chromosome' works in this way. As he put it, 'it was not by studying human nature that linguists discovered the laws of consonant mutation, or Freud the principles of the analysis of dreams, or cultural anthropologists the structure of myths' (Chomsky and Foucault 2006: 6–7). Instead, Foucault conjectured, 'human nature' simply gestures towards an open-ended set of issues that might be examined in a scientific light.

Chomsky replied by suggesting that 'human nature' might constitute an important target of investigation, speculating that we might:

> ask whether the concept of human nature or of innate organizing mechanisms or of intrinsic mental schematism or whatever we want to call it, I don't see much difference between them, but let's call it human nature for shorthand, might not provide for biology the next peak to try to scale after having—at least in the minds of biologists, though one might perhaps question this—already answered to the satisfaction of some the question of what is life. (Chomsky and Foucault 2006: 7)

When Chomsky defends the utility of 'human nature' as a target for scientific investigation in these comments, he is doing two quite distinct things at once. On the one hand, he is moving on from his belief that the ability to acquire language is innate, to encourage scientists to ask whether other capacities might be innate, too. At the same time, he is expressing a belief that the biological and physical sciences might come to explain a variety of human capacities: Chomsky suggests we might move beyond a biological explanation of what life is, and towards a biological explanation of specifically human abilities. These functions for 'human nature' are very different, and it is not especially helpful to run them together. Evidently, one might aim to explain in biological terms—by reference to neuronal selection processes, for example (Edelman 1987)— the possession of cognitive capacities that are not innate.

The question of what is meant by 'innateness' is itself fraught (Griffiths 2002, Mameli 2008b, Mameli and Bateson 2006, 2011). In the context of Chomsky's work on language acquisition, 'innate' seems to mean 'not learned'. Just before the passage quoted above, Chomsky speaks of what is 'biologically given, unchangeable', suggesting that perhaps he wishes to equate innateness with a form of developmental robustness, or canalization (Ariew 1996, 1999). To deny that a trait develops through learning does not amount to asserting that it is developmentally robust: scars are not learned, but their appearance is developmentally fragile. Moreover, to say that a trait is developmentally robust is not to deny that enculturation contributes to its development: it is simply to say that in practice its development is hard to perturb. So, Chomsky's phrasing here is confusing: to advocate 'human nature' as an explanatory goal of the sciences is merely, for Chomsky, to advocate that the sciences seek to explain how diverse human abilities arise as embryos grow to become adults. Chomsky's response seems to be more an endorsement of Foucault's original point than a counter-claim. His rather confusing comments hardly show that the sciences are in possession of a meaty, or restrictive, account of 'human nature'.

Pascal Boyer (2001) provides us with a second example of an appeal to human nature on the part of a serious scientist, but, again, it turns out on inspection that this appeal is compatible with the libertine notion defended in Chapter 4. He aims to explain, among other things, the pervasive appearance of religious ideas in many different human societies. He thinks of these ideas as a by-product of inferential tendencies—regarding the nature of agency, for example—that have been a more or less stable feature of our species' recent evolutionary history. Boyer's scientific theorizing is framed in a manner that explicitly links it to human nature:

There were no solutions to these puzzles [about the ubiquity of religion] until anthropologists started taking more seriously the fact that humans are *by nature* a social species. What this means is that we are not just individuals thrown together in social groups, trying to cope with the problems this creates. We have sophisticated mental equipment, in the form of special emotions and special ways of thinking, that is designed for social life . . . Many animal species have complex social arrangements, but each species has specific dispositions that make its particular arrangements possible. (2001: 27, emphasis in original)

Here, Boyer's gesture to what humans are like 'by nature' involves pointing to a set of robust cognitive dispositions that have been honed by processes of long duration, with the result that they are well adapted to social life. This is all that Boyer's theorizing requires. He is not committed to any further claims about, for example, the development of these robust tendencies being independent of cultural influence. His theorizing also does not require—although Boyer himself sometimes suggests that it does—that the evolutionary historical processes that have shaped these adaptive tendencies consist exclusively of natural selection acting on genetic variation. It would not matter to Boyer's hypotheses about the ubiquity of religion if it turned out that the typical human inferential practices he relies on for explanation turn out to be learned, or if it turns out that their adaptive features are results of cultural evolutionary processes.

In Chapter 4 we examined Cecilia Heyes's (2012a) suggestion that many robust cognitive features of our own species—such as our capacity to imitate others—are the developmental products of learning. Heyes does not mean to imply that these features of our cognitive make-up are easy to alter, or that they are poorly fitted to human social life. She does not even deny that they are products of processes of long duration. Social learning capacities of various kinds develop as a result of prior learning, but this learning takes place in a socially and technically structured environment. Recall her conjecture that the use of technical artefacts such as mirrors helps to explain how infants are able to develop associations between what actions *feel* like as they are performed, and what actions *look* like as they are observed. Recall also her suggestion that these associations can be learned when group behaviours are prompted by a common cause: here, there will be an appropriate correlation between what the actions of others look like as they are witnessed, and what one's own similar actions feel like as they are performed. The social and technical structures that allow imitative capacities to develop are themselves the results of gradual refinement over many generations. It is indeed a mistake to oppose human nature and human culture, but none of this matters to the likes of Boyer because his invocation of the human nature concept is so casual.

5.3 Where Human Nature Meets Evolutionary Research

Scientists like Boyer and Chomsky mention 'human nature' in a positive manner, but that fact is compatible with a bland appeal to the libertine account of human nature underlying their comments. When we turn to models of gene/culture co-evolution, it seems that a distinction between nature and culture plays a more important role in evolutionary theorizing. Co-evolutionary work often makes use of a series of related distinctions between cultural selection and natural selection, between cultural inheritance and genetic inheritance, and between cultural traits and genetic traits. Laland et al. comment that 'Cultural evolutionists tend to view natural selection and cultural evolution as providing competing ultimate explanations' (2011: 1515). Holden and Mace tell us bluntly that 'lactase persistence is a genetic trait, whereas pastoralism and milk drinking are cultural traits' (1997: 604). All of this talk is usually couched in terms of the different influences of cultural and genetic 'inheritance channels'.

Perhaps it is for these reasons that Machery suggests that what he calls the 'behavioural sciences' are committed to something very much like his nomological account of human nature: 'Only those traits that are appropriate subjects of ultimate explanations that appeal to organic evolutionary processes are constitutive of human nature' (2012: 477). If these sciences distinguish organic from cultural selection processes, as they appear to do, then these sciences also appear committed to a distinction between traits that are parts of our nature and traits that are parts of our culture. Once again, it is precisely because many theorists in social anthropology, including Maurice Bloch and Tim Ingold, have been sceptical of the distinction between nature and culture that they have rejected cultural evolutionary work. They detect the same commitments in that work as Machery does.

In what follows I show that these commitments are illusory. There is no objectionable nature/culture distinction at play in the basic business of co-evolutionary work. Moreover, the success of that work provides no support for Machery's nomological account of human nature.

5.4 Developmental Systems Theory (DST)

Some theorists—especially those attracted to DST—have long been concerned about the defensibility of images of evolutionary processes that draw strict distinctions between cultural and genetic inheritance (e.g. Ingold 1995, Griffiths and Gray 2001). Developmental systems theorists focus on the causal cycles through which offspring grow so as to resemble (to a reasonable degree) their parents. Any full explanation of these phenomena of resemblance evidently requires a catalogue of, first, the ways in which genetic, epigenetic, environmental, and cultural resources interact with each other, and second, of the ways in which these temporally extended patterns of interaction exhibit forms of robustness, such that if elements are missing, or misplaced, reliable development need not be wholly derailed. If we hold all these complex developmental processes fixed, allowing only genetic resources to vary, then this variation may result in systematic variation in phenotypic outcome in adulthood. In this sense, there is nothing wrong with talk of the causal influence of genes on phenotypes. But this notion of causal influence is not restricted to genes. If we hold developmental processes fixed, allowing only the sports practised in early childhood to vary, then this variation, too, will result in systematic variation in phenotypic outcome in adulthood.

Stuart Broad is a good cricketer, as was his father Chris Broad. Both have played test matches for England. Suppose we ask the question 'Did Stuart Broad inherit his cricketing talents via the genetic channel, or via the cultural channel?' The proponent of DST says that this is a bad question to ask, and on numerous grounds. First, she will point out the suspicious dualism involved in assuming that the genetic/cultural distinction is exhaustive: development may indeed involve interactions between genetic and cultural resources, but our taxonomy must also make room for the role of resources that are neither straightforwardly genetic, nor cultural, such as chromatin marking, various extranuclear cellular structures, symbiotic gut bacteria, and so forth.

This much will be shared with those who advocate multichannel views of inheritance, and who see more channels at work than merely the genetic and the cultural (e.g. Sterelny et al. 1996). But the DST theorist objects to 'channel talk' in general. Her view is not that Stuart Broad's talents are partly inherited through the genetic channel, partly through the cultural channel, and partly through various epigenetic channels. Instead, she insists that what effects genes have depend on their overall developmental context. That context includes the cultural.

Culture affects nutrition, for example, and that in turn affects how genes are expressed. Conversely, the effect of culture on an individual is equally dependent on overall developmental context, including genetic context. True, variation in both genetic and cultural features can result in disturbances to parent/offspring resemblance with respect to sporting ability. The fact that such disturbances are sometimes reasonably predictable is a result of the stability of the entire developmental system, which acts as a fixed context for local variation in single sets of developmental resources. We cannot demarcate cultural and genetic 'channels', because the very same robust and integrated developmental system is what enables variation in any specific set of resources—be they cultural, cytoplasmic, genetic, or whatever—to have predictable phenotypic effects.

Griffiths and Gray summarize these worries with a useful example:

A developmental systems conceptualisation is more heuristically valuable than a multiple channel or multiple replicator model . . . The idea of dual (or multiple) inheritance systems runs a . . . risk of pushing context dependency into the background. Consider, for example, the methylation inheritance system. The developmental significance of a methylation pattern depends on the gene whose transcription it modifies, and on much else. It is, of course, possible to identify predictive relationships between patterns of methylation and developmental outcomes. However, the idea that these developmental outcomes are transmitted down the methylation inheritance channel obscures the way in which the relationship between methylation pattern and outcome depends on what is happening in numerous 'other' channels. (2001: 198–9)

Oyama generalizes the same point:

An interactionist impulse toward inclusion and parity can be indulged in more or less restricted ways. One may include both biology and culture in one's scheme and make the obligatory antigenetic determinist noises, but to go on funnelling them through different channels of causal influence and inheritance is to lack the full courage of one's interactionist convictions. (2000: 201)

Developmental systems theorists are right to remind us that methylation patterns on DNA, for example, only bring about predictable downstream effects by virtue of their location within structured developmental systems. There is evidently no way to separate genetic and cultural inheritance systems, if that is supposed to involve prising humans apart such that a functional genetic inheritance system might exist in one corner of the laboratory, while a functional cultural inheritance system exists in another. But gene/culture co-evolution and dual-inheritance theories do not need to conceive of the relationship between genetic, epigenetic, and cultural influence in such objectionable ways.

The DST critique leaves intact the possibility of understanding selection as a process that works at a variety of levels of resolution, and it also leaves intact the notion that selection of phenotypic differences may sometimes be driven by genetic differences, sometimes by cultural differences. This is explicitly acknowledged by DST's own advocates.

Griffiths and Gray (1997) argue that various forms of pragmatic simpli-fication, or forms of attentional focusing on parts of the complex devel-opmental system, are legitimate. One can 'study the evolution of particular elements of the life-cycle while assuming that many other elements play a more or less constant role in reconstructing the life-cycle over the relevant stretches of evolutionary time' (1997: 488; see also Sterelny and Griffiths 1999: 108). This all means, for reasons we will shortly examine in more detail, that DST's critique of 'channel talk' need not be fatal to dual-inheritance theory and gene/culture co-evolutionary modelling.

5.5 Lactose Tolerance

In what is perhaps the best-documented piece of work on gene/culture co-evolution, Holden and Mace (1997) have argued that a cultural innovation, namely the development and wide adoption of dairying, brought with it a physiological change in the ability to digest lactose. They argue that once pastoralism became widely established through cultural transmission, it set up a selection pressure favouring lactose tolerance, which consequently promoted the spread of the genes that contributed to that capacity. Subsequent work has emphasized the co-evolutionary nature of this process: lactose tolerance evolves to take advantage of the calories made available by dairying, and dairying is adopted by populations with lactose-tolerant individuals (Itan et al. 2009).

The developmental systems theorist would be entirely justified in resisting any sort of vulgar assignment of pastoralism to a container marked 'culture', and lactose tolerance to a container marked 'genetic'. Dairying is a complex skill, reliant on the availability of appropriate animals, the material resources required to keep them healthy, and a cultivated understanding of how to raise, care for, and milk them. The elements that contribute to the reproduction of dairying practices are certainly not exclusively 'cultural', for effective dairying evidently requires various forms of physiological strength and coordination when it comes to caring for and milking the animals.

It would be equally foolish to cast lactose tolerance as a trait whose reproduction is entirely independent of cultural factors. Humans are unusual not because babies are lactose tolerant—offspring of mammalian species typically suckle in infancy—but because lactose tolerance persists into adulthood. There are genes that causally contribute to this sustained incidence of lactose tolerance by maintaining the activity of the enzyme lactase into later life. That does not rule out cultural factors playing a role in the development of lactose tolerance.

Most obviously, the retention of lactose tolerance in later life requires infant survival later into life, and that in turn requires an enormous array of cultural resources. That is a very broad role for culture. But some individuals take lactase supplements: such people tolerate lactose, but not because they have genes typically associated with lactose tolerance. A recent clinical study by Almeida et al. (2012) suggests that the consumption of certain probiotic yoghurts can improve lactose digestion in otherwise lactose-intolerant individuals, again making room for cultural influence over the development and persistence of lactose-tolerant phenotypes.

More importantly for present purposes, Itan et al. (2010) established that known genetic variants associated with the persistence of lactase into adulthood were unable to account fully for the global incidence of the lactase-persistence phenotype. The authors suggest that this is probably because we have not yet identified all the lactase-persistence genetic variants, but one might just as well conjecture that various epigenetic effects may also help to explain the incidence of lactase persistence. They note themselves that gut trauma, such as gastroenteritis, can result in loss of lactase, and cultural influence over diet can evidently bring about such traumas. Environmental effects can disturb lactose tolerance in other ways. Stress, for example, can result in individuals who are heterozygous for genetic variants that normally result in lactase persistence experiencing lactose intolerance instead (Swallow 2003). Swallow also reports that changes in colonic flora—which can in turn be influenced by cultural nutritional practices—may influence lactose tolerance, and he notes apparent epigenetic modulation of lactase expression. Ruth Mace herself, in a reappraisal of her earlier work, has suggested that gut flora may explain lactose tolerance in Somali nomads who lack alleles associated with lactase persistence (Ingram et al. 2009), and that in some other areas of Africa a comparatively low incidence of lactose tolerance may be explained by the adoption of techniques for processing milk that reduce the advantage of lactase persistence (Mace 2010). These results suggest that cultural context—with respect to diet and stress—and perhaps various other epigenetic influences, are indeed important in the development and maintenance of the lactase-persistent phenotype.

These complications rightly undermine our confidence in thinking of lactose tolerance as purely 'genetic', and dairying as purely 'cultural'. They are also likely to undermine our confidence in thinking that dairying itself is a trait that is untouched by natural selection: even if it is learned, it may be reliably inherited by offspring, and it may have positive effects on survival and reproduction. Taking 'channel talk' too literally can blind us to these complexities.

Crucially, though, none of this undermines Holden and Mace's basic gene/culture co-evolutionary story for the origins of lactose tolerance in humans. It makes perfect sense to distinguish between, on the one hand, the process of reproduction from parent to offspring that allows the relatively slow natural selection of lactose tolerance associated with underlying genetic variation, and, on the other hand, the spread of dairying practices among unrelated individuals, which can establish, in a comparatively swift manner, a set of selection pressures favouring lactose tolerance. The primary explanatory content of this canonical co-evolutionary narrative is immune, perhaps surprisingly, to worries about the difficulties inherent in distinguishing the natural from the cultural.

Developmental systems theorists are not only sceptical of 'channel talk'; they are also sceptical of the idea that functional responsibility for assuring parent/offspring resemblance can be credited solely to genes, or indeed to any small subset of developmental resources. This means that—unlike some working with co-evolutionary models—they have tended to be sceptical of the replicator concept (Griffiths and Gray 1997). They focus instead on lineages of resembling life cycles, within which transgenerational similarity requires (and is explained by) the recruitment of a variety of different elements that conspire in development, and which show various forms of robustness with respect to outcomes that are themselves the product of their systemic organization.

DST does not deny that relatively swift cycles of changes in the behaviours associated with milking domesticated animals can be nested within relatively slower cycles of changes in physiological capacities associated with the digestion of lactose. DST does not deny that while the cyclical processes associated with the reproduction of lactose tolerance typically follow vertical lines of inheritance, the cyclical processes associated with the reproduction of dairying are not so constrained. These sorts of claims are all that is required if we are to conjecture that dairying is transmitted via a cultural channel, lactose tolerance is transmitted via a genetic channel, and the latter has evolved as a selective response to the former. To assert these things entails neither that the ability to milk an animal is acquired independently of physiological capacities, nor that the development of lactose tolerance proceeds independently of suitable social conditions being in place.

5.6 Generalizing 'Channel Talk'

The effects on an organism's activity of various genetic, epigenetic, and cultural modifications depend on the context maintained by the entire developmental system. Developmental systems theorists have consequently stressed the danger that 'channel talk' might blinker us to the manner in which stable effects of any broad class of developmental modifications are typically consequences of this background context. The same caveats apply to analyses framed in terms of cultural and genetic 'inheritance systems'. We can acknowledge these caveats, and continue to analyse evolutionary changes using the resources of gene/culture co-evolutionary models. Moreover, we have also seen that the DST critique is compatible with the recognition of multiple nested selection processes, corresponding to multiple nested cycles of reproduction. This means that the DST perspective also allows us to retain the important heuristic perspective afforded by an adaptationist stance on the relationships between different levels of selection, and on the relationships between different channels of inheritance.

Holden and Mace ask how the comparatively rapid cycles by which dairying skills are acquired affect the slower cycles of reproduction that control the spread of lactose tolerance. If, as I have argued, it makes sense to ask these questions, it also makes sense to ask questions of a more general, higher-order form about how horizontal transmission may undermine or enhance selection acting on vertical inheritance. These sorts of questions, which typically involve taking an adaptationist stance on both the shaping of transmission mechanisms and on interactions between them, have characterized much theoretical work in cultural evolutionary theory, and in evolutionary theory more generally, over the past thirty years or so (e.g. Jablonka and Szathmary 1995, Maynard Smith and Szathmary 1995, Jablonka 2001, Jablonka and Lamb 2005). Sterelny (2001), for example, assesses the prospects for accounts of cultural inheritance by examining how different inheritance systems might compete with each other, how they might facilitate adaptation, and so forth.

To ask how the rapid spread of cancer cells might undermine the integrity of the organisms that house them, and to ask how one might expect selection at the level of individual organisms to respond by suppressing this sort of disruptive selection within the organism, entail neither that the activities of cancer cells can be understood independently of the organismic context in which they find themselves, nor that the flourishing of cancer cells is independent of the very same processes that sustain the organism in health. Similarly, while some casual ways of describing the evolution of 'inheritance systems' might give the impression that 'cultural' inheritance is entirely separable from the processes by which 'organic' inheritance take place, it would be a mistake to think that projects of assessing the costs and benefits of different forms of inheritance presuppose that skills can be acquired independently of the action of genetic and epigenetic processes, or that the reliable development of physiological traits under the influence of genes can proceed regardless of the social context in which those genes are located.

5.7 Machery Revisited

It is reasonable to think of the spread of dairying as establishing a series of niches in which lactose tolerance has a selective advantage. This is good news for the defensibility of cultural evolutionary work, but it is not such good news for restrictive accounts of human nature. We cannot produce a general taxonomy of human traits, assigning those that are the result of what Machery calls 'organic selection' to human nature, while those that are the result of cultural evolution are instead assigned to human culture.

It is a truism of evolutionary studies that development is always a result of interactions between genetic and other resources, including sociocultural resources. Whatever 'organic selection' means, it should not mean 'selection, where the fact that offspring grow so as to resemble their parents is purely explained by appeal to genes'. Genes never discharge this explanatory task alone. Instead, 'organic selection' might refer to phenomena of selection mediated by 'vertical' inheritance: in other words, it might refer to any case where offspring resemble parents with respect to traits that promote survival and reproduction. The problem for Machery is that this leaves open the question of which mechanisms—genetic, epigenetic, cultural, or some complex combination of all of them—explain this resemblance. Perhaps Machery means to equate 'organic selection' with a form of natural selection when facilitated by some specifically 'organic' mechanism of inheritance, but it is hard to see what this appeal to the 'organic' might amount to.

To make these problems vivid, consider the widely reported studies from Michael Meaney and colleagues regarding epigenetic inheritance in rats. Meaney's group has suggested—and admittedly their work remains contentious—that when mother rats lick their offspring, this behaviour results in altered methylation patterns on the DNA of the young rats, and that the resulting modulation of gene expression affects how offspring behave in response to stress (Weaver et al. 2004). Further research has suggested that the young rats who have been licked grow up in such a way that they are more disposed than others to lick their own young (Champagne and Meaney 2007).

If Meaney and colleagues are right about this, it would appear that, at least in principle, response to stress in rats can be subject to natural selection, regardless of whether stress response correlates with genetic variation. The mechanism at work here is thoroughly 'organic', in the banal sense of being biochemical: Meaney and colleagues believe that alterations to methyl groups on DNA are wrought by licking behaviour, and that they give rise to different responses to stress. At the same time, this form of inheritance is manifestly mediated by social interactions between parent and offspring.

Suppose, then, that we were to confirm that rats' responses to stress are indeed subject to natural selection, in Darwin's original and central sense of that term (Darwin 1859). These different responses affect rats' abilities to survive and reproduce, and they are reliably inherited in offspring. Suppose, moreover, that we determine that the mechanism by which inheritance works involves a complicated causal chain, by which the mother's licking affects offspring methylation, which in turn affects response to stress, which in turn affects the ongoing disposition to lick. It is hard to see how it could be in the interests of behavioural science to attempt to assign this selection process either to the 'organic' or to the 'cultural', and any effort to do so seems likely to prompt confusion.

5.8 Natural Selection, Genetic Variation, and Innateness

Much of the business of gene/culture co-evolutionary models is not threatened by the difficulties of untangling the genetic and the cultural, even if those difficulties undermine Machery's defence of human nature. The DST critique of 'channel talk' does not show that co-evolutionary models should be abandoned. But it has critical bite, for it alerts us to the blind spots that can be generated if we take 'channel talk' too seriously. Cultural evolutionary work does sometimes present inheritance in mis-leading ways, and the DST perspective can be especially effective in helping us to see this. First, cultural evolutionists are sometimes too quick to assume that what is naturally selected must also be innate. Second, cultural evolutionists sometimes draw the distinction between individual and social learning in problematic ways.

Consider Henrich and Boyd's (1998) explanation for the evolution of the learning disposition that they call 'conformist bias'. In very broad terms, they argue that conformist bias—which they describe as a ten-dency preferentially to adopt the most prevalent cultural variant in a population—has evolved under the action of natural selection. This bias contributes to reproductive fitness, they argue, because it typically allows individuals to adopt (by social learning) whichever behaviour is more suitable in their environment.

Their mathematical model for the evolution of conformist bias explores this idea by positing distinct 'stages', labelled as 'cultural trans-mission, individual learning, migration and natural selection' (Henrich and Boyd 1998: 221). For the explicit purposes of modelling—and not because they think groups really work like this—they assume that individuals first learn from each other (the 'cultural transmission' phase); that they then learn by trial-and-error interaction with their environ-ment (the 'individual learning' phase); that some individuals then leave the population or join it (the 'migration' phase); that individuals then leave offspring that resemble them with respect to their tendencies to learn from each other and from the environment, and that the number of offspring they have depends on whether they have learned appropriate or inappropriate information (the 'selection' phase). The model also assumes that individuals vary with respect to their degree of conformity in learning from others, and with respect to the degree to which they learn from others compared with learning from their natural environments.

The basic explanatory story they defend with this model is unproblematic in its understanding of the relationship between nature and culture: it tells us that some organisms are at a reproductive advantage because of the ways in which they learn from others, and this is what explains why we learn in a conformist manner. Of course, the fact that the model is unproblematic in this respect does not mean it is unproblematic in all respects. In Chapter 6 we will interrogate the value of cultural models, including this one, in detail. But when we turn from the basic model towards some of Henrich and Boyd's comments about it, we do find questionable pronouncements on the nature/culture relationship. They take it that the manner in which individuals are disposed to learn from others (e.g. the strength of their conformist tendency), and how they apportion their cognitive resources between individual and social learning, are controlled by genetic variation. They also seem to assume that these learning biases are innate. Finally, their model suggests a reasonably clear distinction between individual learning and social learning (referred to in the model as 'cultural transmission'). In the light of what we have discovered in the earlier sections of this chapter, we might object to, or at least query, every one of these assumptions.

Meaney's work on epigenetic inheritance in rats reminds us that not all inherited phenotypic variation needs to be explained by underlying genetic variation. Moreover, one should not assume that when psychological mechanisms are subject to selection, they must also be innate, at least not if that means that various forms of social influence are irrelevant in the explanation of their ontogeny. This is a primary lesson of Heyes's (2012a) work on the manner in which the mechanisms by which we learn from others are themselves subject to cultural inheritance. Finally, as I argue in Section 5.9, the distinction between individual and social learning is itself problematic.

5.9 Individual and Social Learning

The central focus of work in cultural evolutionary theory is on the ways in which the ability of humans to learn from each other has affected, and is affected by, the evolution of the species. A consequence of this, exemplified in the model we briefly examined in Section 5.8, is that cultural evolutionists regularly evaluate the relative consequences of dependence on 'individual' versus 'social' learning.

Heyes has presented empirical evidence against the claim that 'social learning' picks out a cognitive mechanism that is wholly distinct from individual learning (Heyes 2012b). Rather, she argues the very same mechanisms of associative learning that explain how an individual learns from its environment also explain how one individual learns from another. This work does little to undermine the way in which cultural evolutionary thinkers use the social/individual contrast, because even if the mechanisms of social and individual learning are not wholly separable, one can still ask about the relative advantages that accrue to those who use those mechanisms to learn from others, or to learn from the environment. Moreover, while Heyes denies that the basic associative capacities underpinning social and individual learning are distinct, she does suggest that considerable 'tuning' has occurred in, for example, attentional mechanisms, with the result that we can reasonably say that humans are adapted for learning from each other. All of this fits nicely with the cultural evolutionary project of asking why this kind of adaptation has come about.

That project is, however, put under more pressure from the perspectives of DST and niche construction. The latter complements the former, for it stresses the ways in which organismic activities—the building of dams by beavers, the tunnelling of earthworms—affect the developmental and selective environments of their own and subsequent generations. Such approaches remind us that even when considering an individual learning by trial and error in splendid isolation, the very environment with which that individual interacts has typically been structured by the social activities of others.

Alex Mesoudi's account of the social/individual learning distinction is representative of how many cultural evolutionists understand that contrast. 'Social learning', he says, refers to situations 'where one individual acquires information from a second individual nongenetically, as a result of exposure to the second individual's behaviour' (2011: 192). Mesoudi is aware, of course, that this definition admits of degrees. Consider the acquisition of food preferences. We can imagine cases of extreme individual learning: members of a population rarely encounter each other except when mating, and they develop food preferences solely through their own trial-and-error efforts. And we can imagine cases at the opposite extreme: juveniles are exposed to repeated and explicit teaching from expert adults regarding which foods to eat, and which to avoid. Many real-life cases will lie in between these extremes. Psychologists think of 'stimulus enhancement' as a process in which (roughly speaking) a learner's attention is drawn to the object or site of the demonstrator's activity, and they think of 'local enhancement' as a process in which the learner is attracted to the individual demonstrator. These forms of socially mediated attraction can then direct the focus of subsequent individual learning (for discussion see Heyes 1994, inter alia).

The cases of stimulus enhancement and local enhancement do not show that the individual/social learning distinction is a bad one, only that, as with more or less all distinctions, there are some tricky cases that lie in a hybrid zone. Here is a different (and imaginary) case, inspired by discussions of developmental 'scaffolding' (e.g. Sterelny 2003, 2010) and niche inheritance (e.g. Odling-Smee et al. 2003), which causes more genuine trouble for that distinction. Suppose adults have cultivated a fairly large area close to a village, and cleared it of poisonous vegetation. As in the case of extreme individual learning, juveniles are again left to find foods for themselves, with no instruction. Because their local environment is fairly safe, this 'fending for oneself' may also be fairly safe. Perhaps they develop tastes for the plants they find close by, and when adult they cultivate the same plants again, and remove others.

Is this a case of social learning? In terms of the cognitive and perceptual mechanisms used by juveniles there is no difference between this and the extreme case of individual learning; at the same time, the collective efforts of a previous generation have clearly made all the difference to what the younger individuals have learned, by affecting the environment in which they learn. If we focus on the cognitive level, and even on the immediate environmental context, then we may characterize this as purely individual learning. There are, by hypothesis, no other individuals around for the juveniles to interact with as they probe their surroundings. But if we focus on the role played by social activities in ensuring inheritance of adaptive food preferences, then we are likely to describe this as a form of social inheritance, or even social transmission, even if we continue to deny that it is strictly social learning. It is partly for these sorts of reasons that some social anthropologists have been sceptical of any distinction between individual and social learning: they see it as a kind of category mistake (Toren 2012).

In a more general way, we must be alert to the ways in which the supposedly natural environments in which individuals act are typically the results of past social activity of earlier generations. The influential anthropologist Philippe Descola notes that Australian Aboriginals have denied that their environment is a 'wilderness', on the grounds that they themselves are aware of the ways in which their own activities have structured their apparently natural surroundings over thousands of years. He quotes a remark from a leader of the Jawoyn in the Northern Territory: 'Nitmiluk national park is not a wilderness . . . , it is a human artefact' (Descola 2013: 35). These important correctives have issued from work in social anthropology, but they are also regularly stressed by proponents of DST and niche construction theory.

There are hints that close attention from cultural evolutionists them-selves to the ecological and social settings of learning is beginning to put pressure on simple ways of drawing the individual/social learning distinction. Wild chimpanzees sometimes use 'sponges' made from chewed up bits of leaf to soak up water, which they can then drink. In Hobaiter et al.'s (2014) study of learning in wild chimpanzees, they note an apparently new tendency among the Sonso chimpanzee community to reuse discarded leaf sponges, rather than make new ones. They suggest, albeit tentatively, that this behaviour might be a result of the increasing likelihood of chimpanzees encountering discarded sponges, and they note that this might be interpreted as 'a kind of social learning, not influenced by direct observations, but akin to local or stimulus enhancement' (2014: 6). Again, though, its affinity with local or stimulus enhancement is rather weak: we might well classify it as 'social learning', but it need not rely in any way on attention to other individuals, and it need involve no distinctive psychological mechanism.

Hobaiter et al. saw that another chimpanzee began to use moss as a sponge, but apparently without observing any other chimpanzee doing the same. Considered from the purely cognitive perspective, then, this looks like another instance of individual learning. That said, the chim-panzee in question made a moss sponge only after reusing an old moss sponge, which another chimpanzee had discarded. The fact of being in a social group was, therefore, important in explaining how this piece of learning occurred, but it is 'social learning' only in the extended sense that its efficacy is reliant on the activities of other chimpanzees.

The empirical argument-sketches put forward by cultural evolutionists can sometimes be challenged because of their understanding of the individual/social distinction. Consider Mesoudi's claim that 'If individual learning were responsible for variation in human behaviour, then we would expect to see a close match between a person's behaviour and the nonsocial ecological conditions in which that person lives, such as climate, terrain, or local animal and plant species' (2011: 12). The strength of this sort of argument depends in part on what we understand by 'individual learning'. In the imaginary case of food preference acquisition that I just described, the socially mediated maintenance of a local environment could conceivably allow populations to maintain considerable cultural inertia in the face of alterations to their physical surroundings, even when the population's members are characterized at the cognitive level only by individual learning.

Cultural evolutionists have little time for the elaborate discussions of the nature of culture that characterize social anthropology; however, it is partly in order to make room for the ways in which socially structured environments contribute to cultural reproduction that this discipline has been influenced by, among other things, Bourdieu's notion of 'habitus' (e.g. Bourdieu 1977), and Vygotskyan concepts of 'scaffolding' (e.g. Vygotsky 1986).

Kim Sterelny puts ideas of this sort to work in his discussions of social transmission and environmental 'engineering' (2012). Consider a child who spends time conducting trial-and-error learning in the vicinity of an accomplished artisan. Perhaps the child does not observe the artisan at all. Even so, she will be able to experiment with discarded raw materials and prototypes; such raw materials are suitable to their final offices, the prototypes are in various states of completion, and the locale in which they are found is likely to be a safe one in which to experiment. Here we have a socially and materially structured environment that greatly eases the inventive burden on an individual learning how to produce tools. Sterelny is amply alert to worries associated with strong distinctions between individual and social learning, but for other cultural evolutionists the adherence to simple models may obscure his insights.

5.10 Philosophies of Nature

The overall result of this chapter is consonant with the useful distinction Peter Godfrey-Smith (2001) draws between 'research programmes' and 'philosophies of nature'. A research programme may advocate all sorts of provisional distortions and simplifications, with the aim of arriving at particular causal-explanatory claims. Co-evolutionary modelling is a research programme. DST aims at a capacious philosophical framework for describing, with a minimum of distortion, the general nature of organic processes: it is a philosophy of nature. Resembling patterns of organic activities are brought about, generation on generation, by the structured conspiracy of diverse processes, in a manner that allows all types of biological resources to be influenced by prior social activity, just as much as social activities are brought about by prior configurations of biological resources. One who holds to such a view is likely to be sceptical of distinctions between nature and culture, is likely to resist any easy carving up of life cycles into distinct inheritance 'channels', and is likely to deny any good distinction between individual learning and social transmission.

These observations do not amount to a fundamental attack on gene/culture co-evolutionary theories, or dual-inheritance theories, because the manner in which explanatory models are constructed in these trad-itions and the ways in which inheritance 'channels' are conceptualized are typically consistent with the general caveats inherent in DST (Griesemer et al. 2005). This is unlikely to come as a surprise to many cultural evolutionists: when O'Brien et al. (2010: 3797) write that 'all human behaviour is biological' they do so to distance themselves from spurious distinctions between the genetic and the learned. As we have seen, the claim that lactose tolerance co-evolved with dairying does not require that dairying is 'cultural' while lactose tolerance is 'natural', it does not require that dairying is inherited through a set of processes that are entirely independent of those by which lactose tolerance is inherited, and it does not require that lactose tolerance is an innate trait controlled wholly by genes while dairying is acquired independently of any genetic influence.

It is now vogueish in both philosophical and anthropological circles to stress the processual nature of the organism; to stress, that is, the manner in which organisms are smeared over time—sustained and developing through ongoing interactions between a bewildering array of cultural, environmental, and biochemical constituents, every one of which is in turn affected by the activities of the others (Toren 2012). Strikingly similar visions of what organisms are, and how organic lineages change over time, have been put forward by Tim Ingold (2013), John Dupré (2012), Paul Griffiths and Russell Gray (1994), and Susan Oyama (1985).

An organism, says Dupré, is a process. Ingold thinks of humans as 'biosocial becomings'. Griffiths, Gray, and Oyama all stress the manner in which the functional organization that we think of as characteristic of organisms emerges through the collectively orchestrating contributions of a wide variety of developmental resources, none of which is helpfully viewed as prefiguring that final organization. In all of these cases, the theorists in question have shown justified scepticism about distinctions between the natural and the cultural, the evolved and the learned, the individual and the social, the naturally selected and the ontogenetically acquired. What we now see is that these genuine worries do a little, but far less than we might have thought, to threaten the practices of explanation in cultural evolutionary theorizing.

6

The Perils of Cultural Models

6.1 Sense and Circularity

A great deal of work done under the banner of cultural evolutionary theory involves the construction of idealized explanatory models. Indeed, the use of such models is often touted as *the* key virtue of the cultural evolutionary approach. It is important, therefore, to understand both generic worries one might have about the application of such models to the cultural domain and specific worries about the construction of particular cultural models.

In this chapter I begin by demonstrating the ubiquity of model building in cultural evolutionary studies, before moving on in Section 6.2 to consider a pair of very general accusations brought against cultural modelling by Tim Ingold (2007). Ingold argues that cultural evolutionary theorists use models in a manner that is circular, and which thereby offers only spurious confirmation to their proposed explanatory hypotheses. He also argues that the processing of ethnographic data that is required to make them suitable for modelling purposes—more specifically the manner in which they must be abstracted from the context in which they were gathered in order to render them suitable for mathematical formalization—somehow undermines the reliability of these data.

If we read Ingold's criticisms in a superficial way, then they both fail. Evidently, this does not mean that cultural modelling is in the clear. In Sections 6.3, 6.4, and 6.5, I show that cultural evolutionary models sometimes offer dubious formalizations of the hypotheses they claim to test, and also that the support the models' assumptions receive from direct experimental work can be exaggerated. Close attention to specific examples of cultural model building shows that Ingold's concerns—or at least concerns very like Ingold's—about circularity and the use of data away from the context of their generation do offer sources of legitimate concern. While both criticisms highlight some of the ways in which cultural modelling can go astray, they should not be understood to undermine the practice of cultural modelling in general. Instead, they lead on to constructive suggestions for how to use cultural models more persuasively. We need to use these models, because there is no other means by which we can explore hypotheses about the manner in which populational cultural patterns are produced by the aggregated effects of individual interactions.

6.2 Cultural Evolution is a Modelling Science

A dominant tradition of research on cultural evolution—the type of work initiated by Cavalli-Sforza and Feldman (1981) and Boyd and Richerson (1985), and since taken up with enthusiasm by these workers' collaborators and students—proposes theoretical populational modelling as its primary methodological tool. As I argued in Chapter 1, the reason why these approaches to culture are labelled 'evolutionary' is not because their proponents appeal to notions of selection, replication, and so forth at the cultural level. Sometimes they accept such notions; sometimes they deliberately reject them or set them to one side. Instead, they are labelled 'evolutionary' because their proponents see phenomena of cultural change and cultural stasis as arising from interactions between individual humans, and they take it that any attempt to understand such phenomena of aggregation requires the building of models. It is more perspicuous to think of them as kinetic theories of culture.

In the first place, then, cultural evolutionary theorists are convinced that, rather in the manner that speciation and adaptation are populational phenomena that arise through summed events in the lives of individual organisms, so cultural change and cultural stasis are populational phenomena that arise through summed events in the lives of individual humans. In the second place, these theorists assume that many of the statistical modelling practices used within population genetics—but also within epidemiology, ecology, and other modelling sciences—can be profitably adapted to understanding these cultural phenomena.

Henrich and Boyd (2002: 87) tell us that 'Formal models of cultural evolution analyse how cognitive processes combine with social interaction to generate the distributions and dynamics of "representations".' In other words, cultural evolutionary models aim to find, in whatever manner is appropriate, a suitable populational representation of what happens when individuals learn from each other.

It is easy to find practical examples of this commitment to a generic modelling strategy within the community of cultural evolutionists. Consider Joseph Henrich's (2001) exploration of technology adoption and its grounding in human psychology. His explanatory target is the ubiquitous S-shaped curve that describes the rate at which cultural innovations of all kinds are adopted in populations all over the world. New technologies and techniques are accepted slowly at first, their rate of adoption then increases, and finally the adoption rate tails off again as the novelty in question becomes widely used. Henrich asks what sort of individual-level learning processes might give rise to these S-curves. The method he uses to answer that question is to construct a series of theoretical models. We will look at Henrich's reasoning in detail in Section 6.4; for the moment, a rough-and-ready characterization of his approach will suffice. He contrasts two hypotheses for how learning works: 'environmental learning', whereby individuals try out various techniques and adopt the ones that offer them the greatest pay-off relative to their goals; and 'social learning', whereby individuals learn by observing others. He argues that 'environmental learning' typically gives rise to a curve that has the wrong shape—an r-shaped curve, rather than an S-shaped curve—while social learning, and more particularly social learning mediated by 'conformist bias', gives rise not only to an S-curve, but to an S-curve with an appropriate form of early 'lag' that can be seen in empirical data on technology adoption. 'Conformist bias' names a type of social learning that is especially receptive to practices exhibited at high frequency among those with whom one interacts. Moreover, Henrich argues that this 'conformist bias' is a well-confirmed feature of human psychology.

What is offered as a novel insight here is not the claim that innovation follows an S-shaped curve within populations, nor that individuals have a tendency to conform. The novel claim, which is arrived at using the model, is that conformist bias plausibly *explains* S-curves for the adoption of innovations. More generally, the manner in which cultural evolutionists have defended the explanatory pay-off of their approach leans heavily on the notion that it is not immediately obvious what sorts of populational patterns will be produced when learning individuals interact: cultural models are required, and they are informative, by virtue of the fact that 'imitation can lead to unanticipated population-level effects' (McElreath et al. 2005: 485). Likewise, Richerson and Boyd defend the utility of their modelling approach by denying that 'we are all good intuitive population thinkers' (2005: 97): formal models are required to correct and discipline our fallible intuitions.

Wybo Houkes (2012) has pointed out that this purely populational account of the business of cultural evolutionary theory misses out a further ubiquitous concern of its proponents with phenomena of *cumulative* cultural change. Houkes is right about this, but it would be a mistake to think that this means that cultural evolutionists must thereby commit themselves to the centrality of concepts such as cultural fitness or cultural selection. Henrich and Boyd (1998) give a useful schematic account of the sorts of questions that interest cultural evolutionists, which places issues about cumulative evolution in the foreground. As they put it:

1. At a cultural evolutionary level, we want to know how beliefs and values are transmitted among individuals, and why this process generates and maintains differences among groups.
2. At the genetic evolutionary level, we want to understand the conditions under which natural selection could favour the psychological mechanisms posited to explain the cultural evolution of groups. (1998: 217)

In other words, the cultural evolutionist can use models to ask whether and how various forms of learning might sustain differences between groups, and how such forms of learning might promote innovation within groups. Many cultural evolutionary models aim to show that conformist bias is involved in both phenomena. This requires that we show that when individuals learn from each other in a 'conformist' manner, differences between the groups they form will tend to be sustained, and that conformist tendencies allow the preservation of valuable practices at the group level. The sort of modelling work that aims to discharge Henrich and Boyd's first aim need not rely on any notion of cultural selection or cultural fitness, and in the model they put forward in their 1998 paper neither concept is used. Their second aim, which addresses the manner in which, if variation in conformist bias is associated with genetic variation, this psychological disposition might be selected, focuses in their 1998 model entirely on the contribution conformity might make to individual survival, and hence to natural selection in the 'standard' sense according to which it acts on genetic variation as a function of contributions to survival and reproduction. So, while we need to add to our initial characterization of cultural evolutionary theory that it often focuses on the sorts of psychological traits that enable cumulative cultural change, and also on the ways in which such psychological traits may have emerged under the action of selection, we can retain the important point that while both tasks employ populational models, neither task requires that these models adopt notions of cultural fitness, cultural selection, and so forth.

Having established that much cultural evolutionary work proceeds by the use of formal models, and having established that these models need not describe cultural change in a manner that makes it strongly analogous to genetic change, we can also see that one of the most pressing tasks in the evaluation of cultural evolutionary theory is to ask about the nature and reliability of its models, rather than to ask (as many have done) about the senses in which cultural change is similar to organic evolutionary change. The integrity of the cultural evolutionary project is not at all reliant on cultural and organic change being similar; it is entirely reliant on the defensibility of cultural modelling practices.

6.3 Ingold on Cultural Modelling

Tim Ingold has criticized cultural evolutionary theory in a variety of articles over the past twenty years or so. Many of his pieces are polemical, with the result that it can be hard to evaluate their central arguments. Here I consider two criticisms he brings against cultural evolutionary modelling practices. At face value they do not succeed. But a closer investigation of the perils of cultural evolutionary modelling shows that concerns very much like Ingold's—charity suggests that maybe they *are* Ingold's concerns—are justified after all.

The first of Ingold's complaints is the hardest to give credit to. In 2006 Alex Mesoudi, Andrew Whiten, and Kevin Laland produced a manifesto of sorts for cultural evolutionary theory (Mesoudi et al. 2006). Ingold attacked their paper, and in a section of his response entitled 'The Circle of Data and Theory' he alleges that cultural evolutionary models 'have been endlessly self-confirming due to the fundamental circularity of the underlying theory' (Ingold 2007: 15). Towards the end of this chapter

I will argue that some forms of confirmation derived from cultural evolutionary models are indeed circular, and therefore spurious, but Ingold's own reason for declaring circularity is misleading. He takes it that a form of vacuity is present in the way that the entire project of cultural evolutionary explanation is formulated:

> Mesoudi et al. define culture as transmitted information (ideas, knowledge, beliefs, values, skills, attitudes) that affects the behaviour of individuals. They then go on to announce that there is 'ample evidence that culture plays a powerful role in determining human behaviour and cognition' (331). Culture is anything that determines what humans think and do, ergo what humans think and do is determined by human culture! (Ingold 2007: 16)

Ingold here construes the cultural evolutionary project as an effort to secure the claim that culture explains what humans think and do, and he argues that this turns out not to be a substantial claim apt for empirical investigation, but a definitional artefact of cultural evolutionists' decisions to identify culture with whatever factors determine what humans think and do.

Mesoudi and collaborators are not guilty of spinning around the tight definitional circle that Ingold discerns. They endorse Boyd and Richerson's definition of culture as 'information capable of affecting individuals' behavior that they acquire from other members of their species through teaching, imitation, and other forms of social learning' (Mesoudi et al. 2006: 331). As we saw in Chapter 3, this definition is meant to distinguish cultural explanations of behaviour, which under this account appeal to social learning, from explanations that instead appeal to individual learning, or to innately possessed skills. Culture is (for Mesoudi et al.) not whatever explains human behaviour; rather, it is one of many potential sources of information that affects behaviour. The question for cultural evolutionists is to determine when social learning is the explanation for behavioural features of populations, what characteristics social learning has in humans, and why evolution has made us the sorts of creatures that are able to learn from each other in these ways. Such issues have a thoroughly empirical character. They are not settled by definitional stipulation. Ingold's attack misfires.

I do not mean to suggest here that cultural evolutionists' definitions of culture are always unproblematic. We saw in Chapters 3, 4, and 5 that there are clear problems associated with the ways in which social learning is implicitly distinguished from individual learning, and with the more general notion of 'information' deployed in definitions such as the one we have just seen. Even so, a more substantial attack on cultural evolutionary theory will need to look not at what cultural evolutionists say in their framing remarks about the culture concept, but instead at the ways in which an inappropriate view of what culture is undermines the bread and butter of their theoretical and empirical investigations. The models of cultural evolution we have already alluded to in this chapter try to show, for example, that differences between human groups are maintained partly through the action of a psychological disposition to conform. That sort of claim is not an a priori consequence of definition, it is eminently testable, and its standing is wholly independent of what one thinks of the manner in which cultural evolutionists happen to define culture.

Ingold has a second, and more substantial, complaint about the way in which model builders bleach ethnographic data of their content and their context, and in so doing undermine the integrity of those data:

> Anthropologists have amassed a vast corpus of fine-grained ethnographic data, based on long and painstaking fieldwork. These data are, for the most part, qualitative, and are contextually embedded in a way that allows the recovery and interpretation of their meanings. It is in this contextual grounding that their security lies. (Ingold 2007: 16)

He goes on to complain that:

> For Mesoudi et al., however, such grounding is a source of unwanted complexity that has to be discarded in order that data can be transformed into quantities susceptible to 'rigorous mathematical treatments of cultural change inspired by population genetic models' (331). In this process of distillation their security is fatally compromised. Divorced from their original contexts and superimposed upon the 'rich theoretical groundwork for analysing culture in terms of modern evolutionary theory' (331), the data lose any meaning they might once have had. Henceforth their significance is derived from the theory, not from the world. (Ibid.)

The exact nature of Ingold's complaint is not clear. The model builder is likely to respond by agreeing enthusiastically with Ingold that much of the detail and contextual anchoring amassed by ethnographic fieldwork is indeed stripped out when these data are used in the building of models. But, the model builder will add, this is a ubiquitous feature of model building, because it is a necessary one.

A physicist who wishes to understand the behaviour of volumes of gas does not lovingly trace the idiosyncratic biography of each particle contained within that volume. Even if such a thing were possible, explanation of the behaviour of the gas as a whole is achieved by ignoring these details, and instead by assuming that particles act in generic ways for the purposes of modelling by statistical aggregation. Evolutionary biologists, too, abstract away from the details of individual organismic context in the construction of explanatory models in population genetics: perhaps no two individuals run at precisely the same speed; perhaps the contextual effects of running at a particular speed are always contingent on idiosyncratic values for other traits an organism happens to have, and on the peculiarities of environmental circumstance that dictate whether the actual effect of running at a certain speed is to evade a lion, to run into the path of a cheetah, or to break one's leg on a rock. In spite of all this, it is reasonable to think of how the mean fitness assigned to some generically specified trait—fast running—can influence that trait's likely future representation in a population.

Behaviours, skills, and dispositions are inevitably characterized in a comparatively 'thin' manner when they are processed for the purposes of cultural model building. Ethnographic description, on the other hand, tends to aim at considerable 'thickness'.

Typically, the very same behaviour can be correctly described in either thin or thick ways, and there is no one true thickness that we can aim at, independent of our explanatory goals. In Ryle's original discussion of thick description, he asks the question 'What is "Le Penseur" doing?' (Ryle 1971). It is not false to say that he is thinking, even if this is a thin description. It is far more informative to say, for example, that he is attempting to dig his academic department out of a nasty administrative hole by formulating a plausible manner of phrasing his Research Excellence Framework (REF) impact study. While thinking is ubiquitous in human populations, composing REF impact studies is not; and, of course, an observer is only in a position to describe someone correctly as 'composing a REF impact study' if that observer is familiar with local particularities of UK academic culture.

Because cultural models typically aim at the description of very generic processes, it is inevitable that the descriptions they produce will be stated in thin terms. As we have seen, Henrich aims to explain a generic phenomenon, namely the adoption of innovation, by a generic psychological state, namely social learning with conformist bias. A biological model of (let us say) the evolution of industrial melanism in peppered

moths can afford to gloss over the potentially very different developmental routes by which dark pigmentation is produced in different
peppered moths, and even the potentially very different routes by
which alternative industrial pollutants give rise to darkened tree surfaces.
What matters for the basic selectionist story is that melanism is heritable,
and that lighter forms suffer disproportionately from predation by birds.
Similarly, it does not matter for the purposes of Henrich's model if it
turns out that 'conformist bias', say, is a highly abstracted catch-all term
(and therefore a very 'thin' one). So long as people *do* have a disproportionate tendency to imitate more frequent traits, it will not matter if, in
'thicker' terms, these tendencies are a mixed bag. Maybe some conformist individuals are appropriately described as 'keeping an eye on what
neighbouring farmers are up to', while others are described as 'jumping
on the fashion bandwagon'. Just as in biology the goal of describing
developmental processes will sometimes require much 'thicker' description than the goal of describing evolutionary processes, so in the study of
culture the question of how thin a description should be will be dictated
by the nature of the problem one is investigating.

This is a reasonable, if predictable, defence of modelling, but it does
not suffice to dismiss Ingold's concerns. If Ingold is simply worried that
modelling involves thinning out ethnographic data for the purposes of
abstraction, then his worries are misplaced: there is no methodological
sin here to complain about. But close focus on what is required for
cultural evolutionary models to be persuasive suggests that our apparently innocent requirement—namely, that there should be evidence that
individuals have the thinly characterized dispositions that the models
require of them—is sometimes less secure than first meets the eye. Moreover, this lack of security arises from just the sort of decontextualization
that Ingold points to. Finally, once we appreciate that there is still considerable debate over the evidential basis for cultural evolutionary modellers'
claims about (for example) 'conformist bias', we also see that these claims
are sometimes bolstered in ways that are epistemically circular.

6.4 An Ideal for Idealized Models

It is nearly time to look at examples of cultural evolutionary models in
more detail. Let us begin by recalling why cultural evolutionary theorists
get into the modelling business in the first place. Populations of humans

exhibit various patterns over time: they may lose valuable technologies; they may improve valuable technologies; they may sustain norms of punishment, or rites of initiation. These populational features arise at least in part out of interactions between individual humans who make up the populations in question. Ideally, then, we might set about explaining features of human populations by showing how they are brought about by interactions between individuals. The problem, of course, is that these sorts of phenomena cannot be practically explained by tracing the ways in which the life of each human combines in its interactions with others to produce populational patterns. Instead, the modeller attributes general dispositional tendencies to humans, and attempts to find a way of representing these tendencies, which enables their combined effects to be calculated with mathematical tools. A model, when used for this purpose, aims at causal explanation: it aims to show why we see the cultural patterns we do.

A model of this sort needs to clear various hurdles before we are likely to judge it adequate, or persuasive. First, we should be convinced that the phenomena it aims to explain are real: we need to be convinced that innovation really does follow an S-shaped curve, or that technological change generally is cumulative. Second, our models should show as decisively as possible not only that (for example) were a group to be composed of conformist individuals, it would tend to produce S-shaped adoption curves, but also that alternatively characterized individuals would tend not to produce such curves, or that they would be less likely to do so. Third, we should be convinced that the psychological dispositions invoked among humans are genuine: we want empirical evidence that, for example, individuals really harbour forms of conformist bias. After all, if a model tells us that a group of malevolent and powerful gremlins would be highly likely to produce a financial crisis, this hardly shows we should think such gremlins are probably responsible for the financial crisis, unless we have further reason to think such gremlins are real. To the extent that all three hurdles are cleared, we will have reason to think, in this case, that S-shaped curves probably are produced by interacting conformist individuals.

These three observations about models fit well with numerous views about confirmation, and also about the epistemic status of models. Consider the proponent of 'inference to the best explanation'. On the face of things, a model aims to explain some set of phenomena. The explanationist says that to the extent that a model offers the best explanation of those phenomena, we are justified in thinking that it probably latches onto their genuine causes (Lipton 2004). But the explanationist will be concerned about whether some model really shows that a candidate hypothesis can explain the phenomena in question, or whether the model's apparent explanatory virtues are illusory artefacts of idealization. And even if this question is settled, the explanationist will point out that a good explanation needs to appeal to causes of the sort that have some reasonable degree of further confirmation, beyond that accorded by the model in question.

Of course, not everyone is an explanationist. Some proponents of *likelihood* approaches to confirmation, for example, will take it that a better-confirmed hypothesis should have a higher likelihood than its competitors (that is, the hypothesis must confer higher probability on the observations than alternative hypotheses do). In other words, the proponent of likelihood as a measure of confirmation wants to know 'Will a group of conformist individuals be particularly likely, compared with other sorts of individuals, to show an S-shaped curve for technology adoption?'

Many disputes about models, especially about the role of robustness analysis, concern the question of whether this sort of causal-explanatory question can ever be decisively answered with a model (see Weisberg and Reisman 2008, and Odenbaugh and Alexandrova 2011, among others). After all, a model does not contain real conformist learners: in constructing a model various simplifications are required, as well as various modes of formalization that make our representations of learners apt for mathematical treatment. Modellers of conformity and its effects typically assume, for example, that individuals select between just two options (rather than five, or eight, or ten), and that their probability of adoption is a weighted function of the frequency of those options in the population as a whole (rather than a function of the frequency of a subset of options they are aware of). If such a model produces an S-curve, might that result merely be an artefact of some unrealistic aspect of this formalization?

Even if we could somehow be quite sure—perhaps by establishing the robustness of our model's result to changes in parameter values and in the method of formalization—that interacting conformist individuals would indeed produce S-curves, we also want to know what additional evidence we might have that individuals are, in fact, conformist, beyond the mere fact that if they were conformist, their combined interactions would produce S-curves. This is a general problem with views of confirmation that restrict themselves to comparative likelihoods: recall that even if our model tells us that malevolent gremlins would be much more likely than overly risky bankers to produce international financial shocks, we are well short of having reason to believe gremlins are indeed causally responsible for the events of the past decade or so. Proponents of likelihood approaches are typically well aware of these problems (e.g. Sober 1993, 2008), and Bayesians explicitly take account of them by reference to the role of the prior probability of a hypothesis. I suggest, then, that more or less all views about confirmation will tend to converge on our three requirements for models to offer credible causal explanations.

In what remains I will set aside the problem of establishing that the explanatory targets of cultural evolutionary theory are secure. I will assume, in other words, that technology adoption does indeed typically follow an S-curve, and so forth. I do not mean to imply that these sorts of assumptions are always trivial. One might well wonder exactly what is meant by the claim that culture is cumulative, for example. Instead, I will try to show that cultural evolutionary models sometimes fail the remaining tasks assigned for models. Their apparent explanatory results are sometimes artefacts of the modelling strategy chosen, and the psychological assumptions they make are sometimes poorly confirmed empirically. These are, of course, inevitable hazards for any science that makes use of idealized models, and as such they do not show that cultural models should be abandoned. On the contrary, rather than using these problems to fuel a pessimistic attitude to cultural evolutionary theory in general, our awareness of them should act as a spur to produce better models, whose empirical assumptions are more firmly established.

6.5 Perils of Representation

I mentioned near the beginning of this chapter a paper by Henrich (2001), which aims to show that social learning, and not environmental learning, is most probably the causal ground for characteristic S-shaped curves for technology adoption. Henrich's strategy in this paper is a very simple one: he argues that 'environmental learning' is best modelled in such a way that it will always yield an r-shaped curve, no matter what values are chosen for the model's parameters. And he argues that 'social transmission' is best modelled in such a way that it will always yield an S-shaped curve, again regardless of the values chosen for the model's parameters. If environmental learning cannot possibly yield the right kind of curve, and if social learning always does, this gives us good evidence for thinking that social learning underpins the patterns of adoption we observe. Finally, Henrich notes the widely observed 'lag' in the initial adoption of technology, distorting the shape of the 'S'. He claims that if social learning contains an element of 'conformist bias', then suitably shaped curves are produced. Henrich sets all this in the context of a sparse, albeit wise approach to modelling:

> Some readers may criticize this analysis because they realize that a wide variety of mathematical formulations of environmental learning or rational calculations could generate S-curves, and I have not begun to exhaust the possible formulations. That is true. However, merely having equations with the symbols arranged in a particular fashion is not a sufficient riposte. In my view, the trick is to formulate a learning model that is rooted in human psychology, is evolutionarily plausible, is empirically grounded in what we know about human cognition, and still produces S-curves under a wide range of general conditions. (2001: 1007)

In other words, a model offers a plausible guide to causal explanation to the extent that (i) the model's assumptions about how learning works are empirically well grounded, and (ii) the model's ability to produce S-curves is not contingent on any fine choice of parameters. These are indeed sensible recommendations, but they neglect worries about whether the mathematical formalization used in constructing the model can be understood solely as a representation of one form of learning, or whether it is compatible with many different hypotheses about how learning works.

To see why this worry is pressing for Henrich's modelling strategy, we need to begin by asking exactly what he means by 'environmental learning' and 'social learning'. It sounds as though the contrast should be fairly clear: in one case individuals learn from their environments; in the other case they learn from each other. But Henrich's discussion of these apparently distinct hypotheses does not demarcate them neatly, and the resultant ambiguity casts doubt over why he chooses to formalize these hypotheses using one form of mathematical representation over another. His main target under the label of 'environmental learning' is the idea that, in his words, 'individuals acquire and evaluate *payoff-relevant* information about alternative behavioral options by action and interaction in their local social, economic and ecological environments'

(2001: 992, emphasis in original). So, learning from the environment encompasses learning from the social environment. Henrich's contrast, then, is not between learning from the environment and learning from others. Instead, it is between an assumption that 'individuals acquire novel traits by figuring things out, using payoff-relevant information acquired directly from the environment', and an assumption that, instead, humans rely on 'social learning', and more specifically 'biased cultural transmission'.

We still might wonder exactly what this distinction amounts to, especially if environmental learning might include evaluating pay-off-relevant information acquired by observing others. Presumably, Henrich's thought is that some forms of learning involve the adoption of some trait observed in others, without this sort of evaluation. 'Prestige bias' is supposed to be like this: the idea is that individuals emulate whatever it is that prestigious members of their group do, with no need for evaluation of the merits of these individuals' behaviours. 'Conformist bias' is the same: conformists simply adopt whichever behaviours are manifested by high proportions of the population.

The problem with that interpretation is that Henrich's notion of 'biased cultural transmission', which he contrasts with 'environmental learning' encompasses *all* forms of bias, including what he calls 'direct bias'. Henrich explains that 'Direct biases result from cues that arise from the interaction of specific qualities of an idea, belief, practice or value with our social learning psychology. The presence of these cues makes people more (or less) likely to acquire a particular trait.' Henrich then gives as an example 'the practice of purchasing and using cooking oil', which 'spreads rapidly through remote villages . . . because there is something about the behavior or idea that appeals to people' (2001: 997). Under this description of 'direct bias' it would appear that 'direct bias' includes those instances of learning where, for some reason or other, an individual judges a practice to be more attractive than some alternative, given some set of psychological preferences. This sounds exactly like what Henrich understands by 'environmental learning': an individual observes what goes on in his or her environment, notes that some people cook with oil and others do not, and decides that cooking with oil offers some form of benefit that is not present in alternative methods of cooking. It seems, then, as though Henrich's hypotheses have been conflated: 'environmental learning' is (or perhaps sometimes is) a species of social learning under direct bias.

So far, we have merely observed that Henrich's manner of distinguishing 'environmental learning' from 'biased cultural transmission' is confused. We might think that is a minor worry, but it leads to significant problems for his broader modelling strategy. He offers a generic way of modelling biased cultural transmission, which always yields S-curves. He proposes, more specifically, that the basic *replicator equation* offers a suitable formalization of biased cultural transmission:

$$q' = q + q(1 - q)B \qquad (1)$$

Here, q' is the new frequency of individuals with the novel trait in the next time cycle, while q is the frequency of individuals with that trait in the previous time cycle. B simply measures 'the overall difference in the replicatory propensities' of the novel and old traits.

Henrich offers a very different way of modelling environmental learning, which never yields S-curves:

$$q' = p_1 + Lq \qquad (2)$$

Here q' is again the updated frequency of individuals with the novel trait in the new time step, and q is the frequency of individuals with that trait in the old time step. P_1 is the probability that the new trait will be learned from environmental information obtained during that time cycle, and L is the probability that this information is instead inconclusive (and hence that the old trait will be retained). The problem, of course, is that if environmental learning from the behaviours of others *is* a species of direct bias, then environmental learning could just as well be modelled using the replicator equation too. And if that is the case, then environmental learning can produce S-curves. Henrich has not explained why his choice of models is well suited to the processes he aims to understand.

Perhaps I have been unfair to Henrich. Perhaps the distinction he means to draw is between a process whereby individuals adopt behaviours based on what feels intuitively and immediately attractive to them, and a process whereby individuals adopt behaviours based on an estimation of the pay-offs they promise relative to some goal. According to the first model, the use of cooking oil spreads because it simply 'looks right' to users; according to the second model, the use of cooking oil spreads because people ask themselves whether using cooking oil is more likely than using alternative ingredients to promote their health, their social standing, the tastiness of their food, and so forth. We might grant that these really are quite distinct hypotheses about the psychological causes of technology adoption. But we are no closer to understanding why the replicator equation—which can be seen always to yield an S-shaped curve—should be thought suitable for modelling only one of our pair of hypotheses. After all, for users to evaluate cooking oil they need an opportunity to see it in use. As people adopt it, so opportunities for evaluation by others will spread through a population. Whether adoption is guided by reflection on pay-offs or simple intuition does not seem relevant to the issue of whether it is apt for modelling via the replicator equation. The basic problem here is that the environment in which an individual learns can change over time, and with it the individual's ability to evaluate new technologies that are parts of that environment. The environment changes as a function of technology adoption itself. I conclude, then, that, as it stands, Henrich has not given any clear reason why his preferred mathematical model is uniquely suited to a representation of one competing hypothesis over another.

There have been several other occasions when theorists have questioned the apparently secure results of cultural evolutionary models. Consider, for example, the relative roles of so-called 'attraction' and 'selection' in cultural evolution. Sperber's (1996) notion of 'attraction', in his early work at least, is an explanatory abstraction, used to mark the fact that cultural practices are reasonably stable over generations in spite of the fact that individuals do not learn from each other in an especially faithful way. How are we to account for this phenomenon? Sperber's 'attractors' simply record the fact that something or other makes it the case that individual thought patterns tend to cluster around certain forms. For Sperber these facts can relate to widely held modes of cognition, or perhaps to constraints imposed by the nature of commonly encountered problems. When cultural evolutionists instead talk of 'selection', they typically use this term to mark the fact that certain individuals are preferentially used as models for imitation. Sperber's view is that both attraction and selection are important for cultural dynamics. Sperber has also used his views about the nature of attraction to undermine views of cultural evolution that assume strict replication of ideas, in the sense of accurate copying from one individual to the next (Sperber 2000). And yet, a formal model by Henrich and Boyd appears to show that when attraction is strong, the use of models that assume replication in fact becomes more suitable: 'The stronger cognitive attractors are relative to selective cultural transmission, the better the discrete replicator approximation' (2002: 96).

The intuitive pull of Henrich and Boyd's result is nicely illustrated through Claidière and Sperber's (2007) own toy example of smoking. Suppose, for the sake of exposition, that the addictive powers of cigarettes make two patterns of smoking particularly appealing to human psychology: we either abstain altogether, or we are pulled to smoking quite heavily. Henrich and Boyd argue that if these attractors are very strong, they quickly transform a population into one that can be modelled *as though* there is a selection process going on between two replicators, namely abstinence and heavy smoking. If the population begins with a mixture of smoking patterns, smokers above some threshold (say, seven cigarettes a day) will inevitably increase their consumption towards one attractor, while smokers below that threshold will reduce their consumption towards the other. This also means that the pool of models whom new entrants might imitate becomes increasingly polarized. If models are more likely to be emulated according to how close they are to an attractor, then we can quickly see that the population as a whole will soon shift so that all individuals are located either at one attractor or the other. Moreover, the population as a whole will change according to which attractor happens to harbour individuals with the greater tendency to be chosen as models. In spite of the existence of attractors, the dynamics of the process are dictated by the strength of cultural selection acting on discrete replicators.

I will not rehearse the response from Claidière and Sperber (2007) here in detail, but they show quite convincingly that Henrich and Boyd's result is an artefact of the specific manner in which they choose to formalize the relationship between selection and attraction. First, Henrich and Boyd assume that attraction is, in Claidière and Sperber's terms, 'deterministic'. What that means is that individuals above some threshold (e.g. seven cigarettes per day) can only move to one attractor (heavy smoking), while individuals below that threshold can only move to an alternative one (abstinence). Sperber (1996), on the other hand, explicitly formulated attraction such that (for example) even a fairly heavy smoker has a non-negligible probability of attraction towards abstinence. Second, Henrich and Boyd assume that peaks of selection and peaks of attraction coincide. What this means is that individuals occupying the positions of strong attractors are also the most likely to be used as models for emulation. The smoking example illustrates why this assumption is far from trivial: even if heavy smoking is a strong attractor, in the sense that smokers are pulled in that direction as a result of addiction, it does not follow that new entrants to a society are most likely to pick heavy smokers as models to emulate. Perhaps the individuals most likely to be chosen as models smoke only a few cigarettes per day.

These assumptions are what drives Henrich and Boyd's result. The first assumption ensures that when individuals fall within a sphere of attraction they inevitably end up at the attractor in question. If the chance of being used as a cultural model also increases as the attractor is approached, it becomes obvious that after a small amount of time 'old' individuals will be at attractors, and 'new' individuals will determine their own position according to which attractor also happens to have the strongest selective force. In the end, selection is the only force left in play. But once these distorting assumptions are dropped, the basis for arguing that strong attractors are best modelled using replicator approaches falls away.

As a final example of problems associated with the representation of alternative hypotheses about cultural change, consider the very influential work by Henrich and Boyd (1998) on conformist bias, which was also mentioned at the beginning of this chapter. 'Conformist bias' is here understood in a very specific way to imply 'that individuals possess a propensity to preferentially adopt the cultural traits that are most frequent in the population' (1998: 219). *Preferential* adoption here means that if, for example, 70 per cent of people in a population eat with chopsticks, and 30 per cent eat with knives and forks, then it is not merely the case that a conformist is more likely to eat with chopsticks than with knives and forks, but a conformist instead has some probability *greater* than 70 per cent of eating with chopsticks. For this reason, Claidière and his collaborators have proposed that the form of conformist bias that interests Henrich and Boyd, and many other cultural evolutionary theorists, should be referred to as *hyper-conformist bias* (Claidière et al. 2012, Claidière and Whiten 2012). Other researchers have made the same point using slightly different language: Efferson et al. (2008), for example, say that conformity involves an 'exaggerated' tendency to follow the majority.

Henrich and Boyd's (1998) paper argues, on the basis of modelling by simulation, that conformist bias will be favoured by natural selection under a very broad range of environmental conditions. More specifically, they argue that in every scenario where social learning is favoured, conformist bias is favoured too. Again, though, other theorists have been concerned that this result is overly dependent on details of the model: in other words, they have been concerned that it is not robust.

A relatively mild corrective to Henrich and Boyd's model comes from Wakano and Aoki (2007): the corrective is mild because Wakano and Aoki's own model has a very similar structure and makes very similar idealizations to those of Henrich and Boyd. They claim, simply, that Henrich and Boyd's simulation did not involve enough iterations, and that when a similar model is run for a longer period, we see that some conditions where social learning is favoured do not favour strong conformist bias. A far more sceptical treatment can be found in a paper by Eriksson et al. (2007), who question the sensitivity of Henrich and Boyd's results to the nature of the idealizing assumptions made in their model. Their intuitive jumping-off point is the worry that, on the face of things, a population characterized by conformist bias will be highly conservative: if a beneficial novel practice is introduced, then conformist bias will tend to eliminate it. Why, then, think that selection will tend always to favour conformist bias, rather than suppress it?

Eriksson et al. show, quite convincingly, that the result of Henrich and Boyd's model is a consequence of two of their idealizing assumptions, and that once these assumptions are dropped, the selective benefit of conformist bias is no longer apparent. Once again, I do not propose to go into the nature of the dispute between these parties in any detail. The gist is that Henrich and Boyd's simplified models assume choice between just two traits; they also assume that individuals are aware of which traits are available in a population. Evidently both assumptions are unrealistic: there are many more than two ways of building a kayak, and individuals are not typically aware of all the ways there are to build kayaks.

To illustrate why these simplifying assumptions are important for Henrich and Boyd, consider that when only two potential variants are in play, it is more plausible to think that the variant that better promotes fitness will be the more common; hence, that conformist bias (which leads individuals to adopt the more common variant) will itself be more adaptive than alternative modes of learning. It is more plausible to think that in a two-variant population the more effective variant will eventually be represented at above 50 per cent because, since individuals are also assumed to know of these two variants, simple guessing will tend to establish both variants at around 50 per cent, and only a little individual learning is required to tip the representation of the better trait to a level higher than 50 per cent. After that, conformist bias kicks in to promote the trait further. But if individuals are only aware of a small proportion of the potential solutions to a problem, and if the number of potential solutions is very high, one might worry that a population full of conformists will be stuck imitating a poor solution that happens to have become widespread. Eriksson et al.'s alternative models, which relax Henrich and Boyd's assumptions, show many apparently plausible circumstances under which conformist bias is selected against.

Morgan and Laland (2012) have recently noted, in defence of Henrich and Boyd, that Eriksson et al.'s own idealizations may also be unrealistic. This is surely the case, but instead of seeing Eriksson et al.'s work as aiming to show that conformist bias has probably been selected against, we can instead see it as an exercise in scepticism, which renders the adaptive value of conformist bias an open question. They show that Henrich and Boyd's case in favour of conformity's evolutionary significance is driven by the simplifications introduced by their model; hence, that their result lacks a valuable form of robustness. For these purposes it does not matter if Eriksson et al.'s model is also unrealistic.

6.6 Perils of Empirical Confirmation

The foregoing considerations suggest that cultural evolutionary models sometimes offer questionable representations of the processes they seek to depict. But what sort of empirical evidence do we have for the assumptions of these models? What empirical evidence do we have, for example, for the existence of conformist bias? In Section 6.5 I asked the question, 'What reason do we have for thinking that cultural models depict the populational consequences that aggregated collections of individuals with characteristic psychological biases *would* have, *were* such biases to exist?' Here I ask, 'What reason do we have for thinking, on the basis of empirical work, that such psychological biases *do* exist?' I suggest that, at least in the case of conformist bias, the body of evidence is weaker than has sometimes been thought.

Take, as an initial example, a study by Salganik et al. (2006), which Claidière and Whiten (2012: 127) suggest might indicate an instance of conformist bias. Salganik et al. asked what the nature of 'social influence' might be in a simulated online music market. They recruited over 14,000 participants, who were invited first to listen to, and then (if they wished to) to download music tracks by (genuine) bands they had never heard of before. Some participants were simply presented with the music tracks in a random order, and were asked to listen to, and then download, whichever tracks they liked best. They were also asked to rate the quality of the tracks. Others were given the same task, but this time they were given information about the number of times each track had been downloaded by previous participants.

When people had access to this 'social information', it made a difference to the success of tracks. In particular, it undermined the connection between success (in terms of number of downloads) and quality (as measured by evaluations under the experimental treatment where individuals had no access to such information). It also increased the amount of unpredictability, in the sense that it allowed a wider range of different tracks to do well within particular groupings of experimental subjects. So Salganik et al.'s study shows (at least for this experimental population) that 'social information' makes a difference.

One might think this means that conformist bias is at work, but as Olivier Morin (2011) has argued, this interpretation is not licensed by the study. Morin points out that when individuals have access to information about which tracks have been downloaded the most, this does indeed make a difference to the tracks individuals choose to listen to. This is hardly surprising, because (given that the bands are all unknown), there is no other information participants can use to help them decide what to listen to. However, one might think that if conformist bias were strong, then, having settled on a subset of tracks to listen to, individuals would then decide to download the members of that subset with the highest number of previous downloads. But that is not what individuals do. Once the pool of tracks an individual listens to is fixed, download choice continues to reflect quality within that pool (as determined by evaluations in the other wing of the study, where no social information is offered).

Salganik et al. (2006) are right to say, then, that social information makes a difference to market unpredictability. However, the reason why this happens is because individuals download only from among the tracks they listen to, and they only listen to a very small number of available tracks. Social information makes a difference to which tracks they first choose to listen to. Moreover, the early choices of those who first arrive in the study to listen to some small number of tracks will tend to have undue influence over the subsequent tracks people listen to, and in that respect they have unpredictable influence over success as measured in terms of downloads. There is a conformist effect here, in the sense that tracks that have been listened to by lots of other people are more likely to be downloaded by new entrants to the market. But there is no evidence here for any particular psychological mechanism of 'conformist bias', because the conformist effect is merely an artefact of individuals' sampling behaviour when they pick which tracks to listen to.

More generally, it is hard to establish how far conformist bias is supported in the literature because, while many studies have documented important effects involving 'conformity' in some sense or other, the form of conformist bias that is important for cultural evolutionary theory, and which is put to work in its models, takes the very specific form of 'hyper-conformist bias' that we mentioned earlier.

Classic work on conformity in social psychology by Asch (1955), for example, documents the manner in which individual judgements of matters of fact are often deferential to majority views. If an individual is placed in a group that unanimously agrees on an answer (in the classic experimental set-up, it relates to the length of a line), then that individual will often endorse the group verdict even when it is manifestly false. But this does not straightforwardly demonstrate hyper-conformist bias (Efferson et al. 2008). For hyper-conformist bias to be demonstrated, it is not enough to show that when the majority accepts some proposition, this significantly increases the chances that another member of the group will adopt the same proposition. First, that observation is consistent with the thought that individuals form their views by copying the view of a single randomly chosen member of the group. Second, and more significant, consider the basic set-up of Asch's initial experiments. We have a group of seven individuals, six of whom (all confederates of the experimenter) take the view that two lines are of equal length, even though it is fairly obvious that their lengths differ. And suppose we ask the question, 'What is the probability that a seventh individual, who sees the judgements of the others in the group, will make the same false judgement?' For hyper-conformist bias to be demonstrated, it is not enough that this individual has a probability greater than 1/2 of adopting the majority view. Instead, the individual needs a probability greater than 6/7 of adopting that view. Third, Asch found that the conformist effect decays once the group as a whole departs from full unanimity: 'The presence of a supporting partner depleted the majority of much of its power' (Asch 1955: 5). Hyper-conformity, on the other hand, requires that the probability of adopting a common trait remains (for example) above 5/7 when the trait's frequency in the group as a whole is at 5/7, and hence below unanimity.

Other pieces of empirical work that have sought to establish the existence of (hyper-)conformist bias are suggestive, but they fall well short of being conclusive (e.g. McElreath et al. 2005). Efferson et al. (2008), for example, asked people to play a simple game in which they could make money by backing different technologies. The structure of the game allowed them to explore how their volunteers made use of information about others' investment choices. They found that many individuals showed a hyper-conformist tendency, but some ignored the behaviour of others, and one of their subjects even claimed to follow the minority, rather than the majority. They also noted that their data showed nothing about whether conformist behaviour in this particular game was indicative of a stable conformist disposition, or whether the conformist strategy simply emerged over the course of the game as players realized (correctly) that it would lead to high pay-offs. Hence they note, in summary, that 'the issue of how flexible biased social learning is in different settings and over short time scales remains one of the central unanswered empirical questions in the study of cultural transmission' (2008: 63).

Harris and Corriveau (2011) conducted a more recent study on learning in children, where they concluded that children are more choosy about *whom* they learn from, than about *what* they learn. They also argued that children often prefer to learn from individuals who conform. But while this points to some sense in which children prefer conformity in others, it does not show that children themselves are hyper-conformists in the very specific sense required by evolutionary models. Harris and Corriveau's evidence suggests that children often monitor the accuracy of information given by adults, and they defer to those with the better track records. Moreover, in many of the cases where Harris and Corriveau showed that children preferred to learn from conformists, this was in the rather special domain where adults conformed to norms of language use. Here, there is no conceptual gap between getting things right, and doing what others generally do. While a single person in a large group of sincere dissenters might be the only one who knows the best way of building a kayak, a single person in a group of sincere dissenters cannot (so long as the group is a representative one) be the only one who knows the real name for a kayak. The real name for a kayak just is whatever people in those parts usually call it. That raises the possibility that learning from linguistic conformists is an instance of the more general phenomenon of learning from those with better track records of accuracy.

In general, then, it is not possible to co-opt general psychological work on conformity to argue that 'hyper-conformist bias' is itself empirically supported, unless proper care is taken to ensure that the right form of conformity is the focus of the empirical study in question. This point was first noted in early work by Boyd and Richerson themselves (1985: 223), and has been well made in recent papers by Morgan and Laland (2012), Claidière and Whiten (2012), Claidière et al. (2012), and again in the papers mentioned earlier by Efferson et al. (2008) and Eriksson et al. (2007).

After arguing that hyper-conformist bias is poorly supported by existing experimental work, Claidière et al. (2012) and Eriksson et al. (2007) all moved on to design experiments to test directly for hyper-conformist bias in the precise sense assumed by cultural evolutionary models. In both cases they failed to find strong evidence for its existence. Morgan and Laland's (2012) treatment of these issues is far more sympathetic to hyper-conformist bias. They report a recent paper by Morgan et al. (2012), which asked individuals to judge whether two three-dimensional shapes were the same (albeit seen from different angles), or different. They found a flexible form of conformity, suggesting (perhaps unsurprisingly) that when individuals were uncertain of their own abilities, and when they had a large number of other individuals' opinions to make use of, they were indeed disproportionately likely to adopt the majority judgement. These are, of course, just the sorts of circumstances where one would intuitively expect people to make use of the opinions of others. Given the conditional nature of this conformist response, the upshot of Morgan et al.'s experiments for the reality of hyper-conformist bias may seem equivocal: only some people defer to the majority, only some of the time. That said, Morgan et al.'s further analysis also indicates that when individuals make use of 'social information' for judging shapes (i.e. when their decisions are influenced by attention to the views of others), they do so in a manner that is reliably hyper-conformist. If actual decisions do not regularly follow a hyper-conformist pattern, they suggest this results from confident individuals allowing their own perception of shape to outweigh any signal they receive from social information.

So, Morgan et al.'s study does suggest, at least with respect to the sort of task where one can sensibly make use of the judgements of others, that hyper-conformity underlies how we use their opinions. Even so, when it is set against other empirical studies that have failed to find this form of conformity, the overall evidential picture is yet to be settled. More important, the learning strategy used by Morgan et al.'s subjects— namely, learn from others when one is uncertain and the group is large—is obviously a rational one in the circumstances. This means that their data do not support an evolutionary story about the emergence of a specific conformist bias, over an evolutionary story about the emergence of a very general, flexible approach to learning that is manifested as conformity under those circumstances where conformity is wise.

In some cases, it is claimed that theoretical models themselves constitute evidence for the existence of (hyper-)conformist bias. Consider the example of Henrich and Boyd's (2002) paper, where they present a theoretical model suggesting that conformist bias (again, in the sense of 'hyper-conformist bias') can sustain inheritance at the level of populations even when transmission by learning from individual to individual is prone to error. This model is important, because it explores the possibility that cumulative cultural adaptation can proceed at the level of populations in spite of unreliable learning between individuals. When Henrich and Boyd move on in this paper to address the empirical basis for the existence of conformist bias, they comment that 'evolutionary modelling of the cognitive capacity for conformist transmission suggests that genes leading to a conformist bias in social learning will be supported in virtually any environment that also favors social learning' (2002: 101). Here they are referring back to their own (1998) study. In other words, they present a theoretical model showing the potential effects of hyper-conformist bias, and they argue for the reality of this bias by reference to another earlier theoretical model that purports to show such a bias is likely to evolve regardless of the precise nature of social learning. But we addressed this earlier model in Section 6.5, and noted (following Eriksson et al. 2007) that its results appear to be arte-facts of the model's idealizations.

Henrich and Boyd suggest another source of empirical evidence for conformist bias: 'by analysing the dynamics of transmission, Henrich (2001) provides evidence of conformist transmission from field data on the diffusion of innovations' (2002: 101). That is another paper we examined in Section 6.5, and we saw that it, too, uses a questionable theoretical model to argue that 'environmental learning' cannot account for S-shaped curves of innovation adoption, while conformist bias can. I suggest that this form of apparent empirical confirmation is illusory.

6.7 In Praise of Models

We can now return briefly to Ingold's (2007) criticisms of the cultural evolutionary project, and more specifically of the use we find there of models. In broad terms, he complains that these models exhibit a form of objectionable circularity in the relation between theory and data, and he complains that the manner in which thick data are thinned out for the purposes of modelling robs those data of their security. We have now seen that versions of both criticisms (which may also be Ingold's own versions) have some weight. It is epistemically objectionable, and a form of circularity, to argue that contentious empirical assumptions of a theoretical model are bolstered by appealing to the role of those same assumptions in other theoretical models, whose causal-explanatory vir-tues are dubious. If a given model uses implausible idealizations to derive the result that conformist bias is responsible for some phenomenon, one cannot use that result to argue that conformist bias is most likely a real psychological trait when establishing the plausibility of the assumptions of a different theoretical model.

What we can do, of course, is appeal directly to experimental work in search of confirmation of our models' assumptions. Here, the case of conformist bias shows the risks of assuming too quickly that this task can be discharged by data borrowed from other disciplines. As we have seen, the problem is not that 'thin' forms of description are inevitably inferior to 'thick' forms of description. The problem, instead, is that without paying close attention to the precise nature of the manner in which data on 'conformity' have been generated, and without attending to the meaning of 'conformity' in these experimental contexts, we have no guarantee that the phenomena in the studies we consult, whether thinly or thickly described, are instances of conformist bias in the very precise, albeit very thin, hyper-conformist sense required by cultural evolutionary models. As a matter of fact, they sometimes are *not* instances of hyper-conformist bias, even if they may appear to be so at first sight (as in the Salganik et al. 2006 study, and probably in the Harris and Corriveau study from 2011).

Even the most careful treatments of the nature of conformity slip up sometimes in their interpretation of exactly what conformity is. Claidière and Whiten begin their (2012) paper by pointing out that the term 'conformity' has been used in so many ways that it is unclear whether different studies of conformity all focus on the same phenomenon. Their own discussion then struggles to control behavioural and psychological definitions of conformity. Their remarks on 'nonconformity' are explicitly couched in terms of individuals' motivations: 'Nonconforming individuals are simply not sensitive to group pressure; they are not motivated to be similar (conformity) or different (anticonformity), from others.' Contrast this with their suggestion that 'conformity' should be defined 'by the fact that an individual displays a particular behaviour because it is the most frequent the individual witnessed in others' (2012: 126). This says almost nothing about individual motivations that drive such dependencies: an individual with no conformist motivation will be more likely to witness, and, therefore, to learn, those behaviours that happen to be most common in that individual's environment. Such an individual is 'conformist' in the behavioural sense, but 'nonconformist' in the motivational sense. Ingold's generic concerns about what happens when data are shared across disciplines and cut loose from context are well founded.

It might appear that my argument recommends abandoning theoretical cultural modelling, and thereby abandoning a dominant programme of work in cultural evolutionary theory. Far from it. I have not presented arguments showing that cultural modelling is inevitably doomed. Here, I distance myself from the very strong claims to circularity in these areas apparently endorsed by Ingold. Most obviously, reasonable concerns about importing data from one domain into another are best thought of as warnings, which can be overcome so long as we take due care over translation. This is something that Boyd and Richerson, for example, have long been sensitive to (1985: 223). There is no principled barrier to establishing empirically grounded claims about the typical means by which we learn from each other, for example, and the work I mentioned earlier by Morgan et al. (2012) takes seriously the task of giving experimental confirmation to the specific assumptions of Henrich and Boyd's models.

Only slightly less obviously, there remains a valuable role for cultural evolutionary modelling simply because we have no other techniques available to us for answering questions about the relationship between broad populational patterns and the individual learning dispositions that at least partially explain them. If we know about broad demographic facts, it does not follow that we understand what gives rise to them, and if we know about individual psychological dispositions, it does not follow that we understand the kinds of effects they produce in aggregation. The only tools that give us hope of exploring these relations between individual and population-level properties are theoretical models and simulations of various sorts. Here, Henrich and Boyd's justification for their broad approach is entirely correct:

> Formal models help us to develop a more complete understanding of cognition and culture. Our minds are not well equipped to understand the population-level outcomes that arise from numerous minor interactions, weak cognitive biases, random errors, migration rates and micro-level decisions. The history of population ecology, epidemiology and evolutionary biology over the last 40 years provides strong evidence that simple mathematical models provide powerful tools for understanding such problems. (2002: 109)

This leaves open our evaluation of particular cultural models. Indeed, we have now seen enough to understand the scepticism sometimes expressed regarding how such models are used. Even so, the moral of this is not that we should throw models out altogether. Henrich and Boyd's conclusion stands: 'It will never be enough to focus on the mind and ignore the interactions between different minds. To keep track of such interactions some kind of population dynamic models will be necessary' (2002: 110). These models can, in principle, be backed by empirical research that aims to confirm their psychological assumptions.

Some go so far as to argue that robust models themselves offer confirmation of causal hypotheses (e.g. Weisberg and Reisman 2008): they argue, in other words, that if one can show a suitably robust linkage between, say, conformist bias and S-shaped adoption curves, one thereby provides evidence in favour of the proposition that conformist bias causes S-shaped curves. I have not gone so far as to endorse that claim in this chapter, and an evaluation of the general nature of robustness analysis goes beyond what I aim to achieve here.

Regardless of where we stand on this issue, we can agree on many other valuable functions that models can play. They can serve heuristic functions, by pointing to potential hypotheses to test regarding psychological biases. They can be used to cast doubt on hasty claims about cultural change, such as the claim that cumulative evolution can only proceed with faithful learning. Here, modelling is innocent because it is used to criticize a piece of intuitive, informal, and eminently unreliable modelling; namely, the hunch that if individuals learn from each other in error-prone ways, cultural change cannot be cumulative. In achieving this sceptical function they can also establish correlative possibility proofs, for example that cumulative cultural change *might* proceed under conformist bias. Reflections on models, and attempts to construct them, prompt us to ask valuable questions about what we mean in opposing 'individual' or 'environmental' learning on the one hand, and 'social learning' on the other, and they can help us to think about what sort of demographic facts, if any, might discriminate between them. There is no reason to reject cultural models, so long as we are aware of their perils.

7

Populations, People, and Power

7.1 Power and Populations

The kinetic theory of gases aims to explain the large-scale behaviour of volumes of gas in terms of the combined effects of the particles making up that gas. Of course, the kinetic theory does not track and sum the behaviours of each particle in a container. Instead, the kinetic theory makes idealizing assumptions about the nature of every particle in the volume, and then sums these behaviours using statistical tools to derive the behaviour of the gas as a whole. By thinking of the gas as a very large collection of tiny particles bouncing off the walls of containers, and off each other, we can explain (for example) the relationships between the pressure, temperature, and volume of the gas.

We have seen that much cultural evolutionary theory aims to explain cultural phenomena in a rather similar way. Mathematical tools are used to explain phenomena manifested by populations of humans in terms of the aggregated effects of interacting individuals. Once again, these individuals are not each tracked in detail; rather, they are given idealized characterizations, and their behaviours are summed using statistical tools.

Richard Lewontin has long been a sceptic of cultural evolutionary theory, and our understanding of it as a kinetic theory now puts us in a good position to evaluate Lewontin's belief that the theory cannot handle phenomena of power. Roughly speaking, Lewontin worries that by framing cultural change in terms of the aggregated interactions of individual people, cultural evolutionary theory ignores the ways in which cultural changes are subject to forms of centralized control:

no model of cultural evolution of which I am aware takes account of power. The people of Bavaria are predominantly Catholic while Westphalians are Protestant, not because somehow Lutheranism was more appealing to northerners but because at Augsburg in 1555 the warring German princes and the Holy Roman Emperor made peace using the rule of *cuius regio, eius religio*, which allowed rulers to enforce their own religion in their own dominions and to expel those who were recalcitrant. (Lewontin 2005: 3)

An uncharitable reader might take Lewontin to be claiming that a ruler could simply make it the case that his subjects were either Catholic or Lutheran. This surely exaggerates the power of rulers after the Peace of Augsburg. The principle of '*cuius regio, eius religio*' granted them the right to determine the *official* religion of their domains. And yet, this power was limited in a number of ways, which left room for the appeal of different religious orientations to individuals in spite of 'top-down' efforts at control. First, it is not unusual for historians to suggest that the Peace aimed to preserve the religious status quo. Hence some urban magistrates and rulers chose Lutheranism because a substantial group of their subjects were already Lutheran; their subjects did not become Lutheran because of the choice of their rulers. Second, the Peace granted no official recognition to Calvinists. This did not mean that Calvinism disappeared. On the contrary, the number of Calvinists grew, and this resulted in a further Imperial legal accommodation. Third, the power of rulers was patchy: consider the later 'court Calvinism' in Brandenberg, where local pastors and parishes remained stubbornly Lutheran in spite of the official court religion. Individual choices over religious affiliation continued to play a significant role in shaping the theological demography of German lands after the Peace of Augsburg (Rublack 2004).

What I take to be Lewontin's general point is easier to express by means of a simpler example. Musical recordings are the sorts of things that increase and decrease their frequencies in populations. But it does not follow that successful recordings must be especially well suited to the musical preferences of listeners. In Chapter 6 we came across Salganik et al.'s (2006) study of behaviour in simulated music markets. Recall that over 14,000 individuals were given access to a website allowing them to listen to, and download, a range of (real) music tracks from previously unknown bands. The investigators found that giving listeners information about the popularity of these tracks, in terms of the number of downloads by other users of the simulated market, could make a large difference to the subsequent success of those tracks in the market:

Increasing the strength of social influence increased both inequality and unpredictability of success. Success was also only partly determined by quality: The best songs rarely did poorly, and the worst rarely did well, but any other result was possible. (Salganik et al. 2006: 854)

This result highlights the dangers of unreflective appeals to cultural 'fitness' which unwisely link a trait's overall rate of spread with its conformity to local tastes.

Lewontin—this time in collaboration with the historian Joseph Fracchia—makes just this point elsewhere, in another attack on cultural evolutionary theory:

the success or 'selectability' of a solution, whether it will be short- or long-lived, depends not just on its 'fitness' in regard to specific problems, but also and crucially on the amount of power behind it—power that can force its 'selection', at least temporarily, regardless of whether it is a 'fit' solution. (Fracchia and Lewontin 2005: 21–2)

Fracchia and Lewontin are right to be wary of the sort of evolutionary view that casts cultural 'traits' as the sort of things that spread through populations solely in virtue of their suitability with respect to local cultural environments, which in turn are constituted by individual preferences. This sort of conception is invited by drawing close analogies between cultural change and biological adaptation, and the example of music downloads shows how it can mislead. The lesson to draw from Salganik et al. (2006) is that when individuals have access to information about previous downloads, the success of a track often fails to correspond to its appeal to musical preference. Instead, track success can be an artefact of the wholly contingent matter of which tracks early entrants to the market happened to listen to. These early entrants thereby have disproportionate power over the subsequent behaviour of the population as a whole. Fracchia and Lewontin's claim is not that individual preferences are never relevant; their fundamental claim is that inequalities of power are sometimes relevant, and that cultural evolutionary theories struggle to take account of them.

At this point cultural evolutionists might well wonder what problem they are guilty of. The kinetic approach does not assume that cultural traits spread in the same way as organic traits. It certainly does not assume that cultural traits invariably spread according to the perceived utility they offer to each member of the population. Indeed, populational models might be useful if we want to investigate the nature of what Salganik et al. call 'social influence'. We have seen several times in this book that the models examined by cultural evolutionists often explore outcomes that occur in populations where, say, individuals are disposed to imitate the majority ('conformist bias'), or where individuals are disposed to imitate the successful ('prestige bias'). In the case of prestige bias, power is not equal among group members: successful individuals have more influence over the cultural trajectory of the population than others (Henrich and Gil-White 2001). Simple phenomena of power present no special problem for kinetic theories of culture.

7.2 Powerful Actors

The authors of the study on music downloads made a telling remark:

Our experiment is clearly unlike real cultural markets in a number of respects. For example, we expect that social influence in the real world—where marketing, product placement, critical acclaim, and media attention all play important roles—is far stronger than in our experiment. (Salganik et al. 2006: 855)

Here is where Fracchia and Lewontin's concerns about power become more pressing than in the simpler case of 'social influence' directly explored by Salganik et al. A single powerful music distributor might ensure through various contractual agreements, favours owed, and intimations of future treatment that a given song is played everywhere, endorsed by prominent critics, and known to be positively endorsed. Even if people do not like the song much, its ubiquity might ensure that it gets downloaded much more often than a different song that is much 'catchier', but which has little corporate backing.

Phenomena of power have the potential to place control of cultural trajectories in the hands of one or a few actors. This can be especially striking in those cases where the institutional background presupposed by a populational analysis is itself under the control of a handful of agents. That thought is nicely illustrated by an example that occupies much of Geertz's notorious paper 'Thick Description' (1973); namely, sheep stealing. In direct contradiction of Richerson and Boyd's claim that 'Culture is (mostly) information stored in human brains' (2005, 61), Geertz tells us that 'the main source of theoretical muddlement in contemporary anthropology is a view which . . . is right now very widely held—namely, that . . . "culture [is located] in the minds and hearts of men"' (1973: 11). Geertz concedes that to steal a sheep, one must have something 'in one's head'—one must know how to steal a sheep. But this is not sufficient for sheep stealing, or so Geertz urges. One cannot steal a sheep unless one's actions take place in a context of appropriate social institutions: sheep must be regarded as property, their removal from a given location must be held to be a crime, and so forth. Whether it is possible for an individual to steal a sheep, therefore, depends on more than the individual's cognitive make-up. Epidemics of sheep theft may sometimes occur, which could perhaps be modelled from a populational perspective. But an individual cannot steal a sheep unless sheep can be the property of others. Sheep cannot be the property of others in the absence of legal institutions that uphold property rights. Cultural evolutionary models often assume the existence of these institutions as background conditions within which a populational analysis takes place. Such assumptions are reasonable when these background conditions are indeed stable. However, a powerful ruler could instantly eliminate sheep theft, by instantly removing the sorts of institutional safeguards that enable the ownership of sheep. Just as the kinetic theory of gases would fail to work in circumstances where gas molecules are simply moved around en bloc—created, destroyed, or transmuted by a powerful actor—so the kinetic theory of culture fails if the cultural practices it speaks of are subject to the whims of those controlling the social institutions that underpin them. To return to Ryle, whose own work is a foundation for Geertz's reflections, 'A statesman signing his surname to a peace treaty is doing much more than inscribe the seven letters of his surname, but he is not doing many or any more things. He is bringing a war to a close by inscribing the seven letters of his surname' (Ryle 1971: 496).

7.3 Powerful Organizations

Our discussion so far has focused on power as a relation between individual agents. Steven Lukes has argued that such a conception of power is inadequate (Lukes 2005). Power, Lukes argues, can be exerted not only by individual people, but also by interest groups, unions, and so forth. One might worry that cultural evolutionists invariably assume that individual people are the only important loci of power. We might assume, in other words, that the only way that cultural evolutionists can acknowledge power is via concepts like 'prestige bias'. And so we might be tempted to conclude that cultural evolutionary theories are bound to be blind to important organizational loci of power.

Again, this verdict would be unfair. Consider the work of Peter Turchin. In a series of publications over the past decade he has attempted to bring techniques of mathematical modelling to the study of human history. He shares with cultural evolutionists a desire to use idealized models in order to understand broad trends in human social behaviour over time. Like cultural evolutionists, the models he uses are often adapted in various ways from the biological sciences (his own home discipline is ecology). To give just one example, Turchin has built a highly abstract mathematical model that is designed to explain the collapse of states (e.g. Turchin 2009a). This model takes into account not just the number of inhabitants within a state, but also the accumulated resources of the state itself. In calculating those accumulated resources, Turchin introduces a variable that measures the proportion of surplus production collected as taxes: this is, in effect, a measure of the power of the state to extract resources from individuals. He also assumes that the state's resources have an overall effect of reducing conflict between individuals within those states. Here, again, the model explicitly acknowledges that state power (rather than individual power) acts to restrict individual action.

Turchin's work refutes the thought that cultural evolutionists take account of power only when it is exercised by individual people. But models of the sort Turchin deals in have significant limitations (and I doubt that Turchin would deny these limitations). They allow us to explore some of the effects of state power, when power is understood in a highly abstracted way. Lukes, in his own treatment of power, stresses that facts about the procedures governments follow, together with similar facts about the processes of other controlling organizations, can have significant biasing effects. They can ensure that concerns that are important to the interests of marginalized groups go unheard, because the procedures in question impede the recognition—perhaps for bureaucratic reasons—of the issues in question. For these reasons one cannot fully understand power—i.e. how some agents' interests are systematically ignored, while the interests of others tend to predominate—without understanding these procedural facts.

As we have seen, Turchin's model for state collapse introduces a single variable that measures the degree to which the state can acquire surplus production from individuals. Even if this approach tells us something about the typical consequences of one form of institutional power, it does not tell us about the kinds of processes that make it the case that institutions are systematically open to some people's concerns, and systematically closed off to others. To put the distinction crudely, even if we can learn something from Turchin's models about the effects of state power, this does not mean that we also learn about the detailed internal mechanisms whereby institutional power is constituted.

7.4 Accommodating Power

I have raised two broad sets of worries for populational approaches to cultural evolution: can they deal with very powerful individual actors, whose idiosyncratic influence seems to elude aggregative models, and what, if anything, can they tell us about the mechanisms whereby organizations exert power? Cultural evolutionary theories have three options for how to respond to these concerns, which I call restriction, presumption, and reduction. The first two strategies defend comparatively modest roles for the populational approach. *Restriction* concedes that populational models have limits, and it restricts their application to times and places where human societies have been free of interest groups, governments, unions, and perhaps gross asymmetries of individual power. This may not be too damaging to the value of such models in exploring early hominin history, even if it does limit their applicability to modern settings, for it is frequently supposed that bands of hunter-gatherers were largely egalitarian with respect to individuals' access to economic resources (Knauft 1991, Boehm 1999). *Presumption* allows that populational models might be used even when complex organizations and institutions have arisen, but it regards facts about these organizations and institutions as fixed background conditions against which individuals interact, and which constrain individual action. This is a reasonable strategy when it comes to understanding those individual interactions and their long-term effects, but it is a strategy that simply omits to explain many features that are admitted as important in determining cultural change. In other words, this is another strategy that explicitly concedes the explanatory limitations of populational approaches.

Reductionism, on the other hand, says that although governmental structures, judicial procedures, and so forth have significant influence on cultural trajectories, these features themselves can be understood via populational models, because they arise as a consequence of interactions between individuals making up the government, the judiciary, and so forth.

I do not propose to decide between these three options here. Instead I note that, to the extent that one advocates population thinking as the way to understand *all* cultural phenomena, one takes a side in the long-standing debate over the plausibility of methodological individualism in the social sciences. The populational approach claims that if we want to understand cultural change, 'The key is to focus on the details of individual lives' (Richerson and Boyd 2005: 59). By focusing our attention on individual humans and their interactions, this form of thinking tends to draw our attention away from the potential efficacy of bureaucratically structured organizations such as trade guilds, learned societies, schools, councils, governments, and so forth.

I do not mean to imply that kinetic approaches deny that social groups can have very important properties. Many cultural evolutionists regard cultural group selection, for example, as a potent force. That said, their primary way of exploring cultural group selection is via an assessment of the aggregated effects of individual interactions in groups with different compositions. Moreover, cultural evolutionists have devoted considerable attention in recent years to studying the emergence and efficacy of institutions. But they define these, in a manner reminiscent of Lewis's (1969) account of convention, as 'emergent phenomena that arise at the population- or group-level from the individuals' interactions, decisions, and learning. They are first and foremost self-reinforcing, dynamically stable equilibria that arise as individuals' norms converge and complement each other over time' (Richerson and Henrich 2012: 40).

Cultural evolutionists are concerned with giving analyses of broad social institutions such as property (e.g. Skyrms 1996) and marriage (e.g. Henrich et al. 2012), rather than structured social organizations such as unions, universities, and legislative assemblies. Their analyses are explicitly reductionist, in the sense that they aim to show how the stability of institutional practices can result from the aggregated behaviours of individuals. Richerson and Henrich intimate that institutions should only be considered as respectable explanatory devices if they are also reducible to patterns of interaction between individuals: they 'have long lacked sufficient micro-level foundations to be taken seriously by researchers in the economic and evolutionary sciences' (Richerson and Henrich 2012: 40). In these respects they are methodological individualists.

7.5 Holism

Alex Mesoudi rightly diagnoses the poor reception cultural evolutionary models have had in some quarters as a consequence of an unwillingness to explain sociocultural phenomena in ways that reduce the macro-level features of societies to micro-level descriptions referring to psychological states (2011: 52). Fracchia and Lewontin, for example, have complained about the tendency of Boyd and Richerson's work to 'dissolv[e] society into a collection of atomistic individuals' (Fracchia and Lewontin 1999: 69). Here they echo a far earlier sentiment from Kroeber:

The findings of biology as to heredity, mental and physical alike, may then, in fact must be, accepted without reservation. But that therefore civilization can be understood by psychological analysis, or explained by observations or experiments in heredity, or, to revert to a concrete example, that the destiny of nations can be predicted from an analysis of the organic constitution of their members, assumes that society is merely a collection of individuals; that civilization is only an aggregate of psychic activities and not also an entity beyond them; in short, that the social can be wholly resolved into the mental as it is thought this resolves into the physical. (1917: 193)

Mesoudi's explanation of the deeper sources for this resistance is less plausible, and it might lead us to be too dismissive of anti-reductionist approaches. He briefly cites Durkheim, and then suggests that 'this reluctance to reduce cultural phenomena to individual psychological processes may have its roots in the mind body dualism inherent in many of the nonscientific approaches to culture' (2011: 52). This is a mistake. Durkheim's anti-reductionism was based on something akin to a scientifically respectable emergentism, or holism, which regarded the behaviour of wholes as more than the mere aggregated sum of their parts.

The most convincing holist arguments have conceded that organizations and institutions have nothing more than individual persons as their parts. Instead, they have argued that explanatory reductionism is implausible in complex social contexts, where sociocultural phenomena have an autonomy that may arise out of relations between individuals, but which cannot be reduced to claims about individual psychology. A similar style of argument is popular these days among those anti-reductionists who might concede that biological organisms are composed of nothing more than molecules, while denying that the best way to understand the capacities and dispositions of whole organisms is via an understanding of those molecules (e.g. Dupré 1993, Moss 2003).

The organic analogy has been influential within certain schools of social science. Its importance does not lie in the claim that society functions as an organism, with analogues of metabolism, repair, mutually adjusted organs, and so forth. Instead, it reminds us of the rashness of assuming that the operation of a complex whole is always best explained by attending to its micro-level parts. Durkheim put the argument well:

But, it will be said that, since the only elements making up society are individuals, the first origins of sociological phenomena cannot but be psychological. In reasoning thus, it can be established just as easily that organic phenomena may be explained by inorganic phenomena. It is very certain that there are in the living cell only molecules of crude matter. But these molecules are in contact with one another, and this association is the cause of the new phenomena which characterise life, the very germ of which cannot possibly be found in any of the separate elements. A whole is not identical with the sum of its parts. It is something different, and its properties differ from those of its component parts. Association is not, as has sometimes been believed, merely an infertile phenomenon; it is not simply the putting of facts and constituent properties into juxtaposition. (Durkheim 2006: 52)

Needless to say, many have disagreed with Durkheim. Such arguments have constituted much of the ground on which debates regarding methodological individualism have been conducted. I do not wish to defend Durkheim here, but one should not think that in endorsing emergentism one is committed to a form of mind/body dualism.

7.6 Three Kinds of Explanation

Even if one is sceptical of holist, emergentist views of social institutions and organizations, it does not follow that the kinetic approach advocated by many cultural evolutionists is appropriate. Our discussion brings to the fore a useful threefold taxonomy of part–whole explanation, which has been drawn in a compelling form by Levins (1970, inter alia), and applied in various fruitful ways by Wimsatt (e.g. 2006) and Godfrey-Smith (2012). I will call the three approaches functional holism, functional decomposition, and population thinking.

We have already characterized *functional holism*. It denies that the activities of complex wholes—organisms, societies, and so forth—can be understood merely by examining how their parts act in juxtaposition. *Functional decomposition* is explicitly opposed to functional holism. It asserts that the workings of a complex whole are explicable by reference to the combined causal interactions of its parts. Importantly, instances of functional decomposition—say, an explanation of the workings of a car by appeal to the engine, the carburettor, the driveshaft, the differential, etc.—do not appeal to *population thinking* (Godfrey-Smith 2012).

Population thinking explains the whole by reference to the combined interaction of its parts—so it, too, is opposed to holism—but in a way that distinguishes it from functional decomposition. The question of precisely when populational models become appropriate, and when functional decomposition should be used, is bound to be vague. Populational models apply best to wholes with very large numbers of parts, which can all be characterized in very similar ways. Hence the kinetic theory of gases is exemplary of this mode of thinking. That theory assumes that there are no explanatorily relevant differences between individual gas molecules, and that there are very many of these molecules. Other populational models in evolutionary game theory or evolutionary economics might have just a few types of sub-part, such as hawks and doves, innovators and imitators, and so forth. At the very least, the number of token parts must greatly outstrip the number of part types. One need not be a holist, then, regarding the relationship between individual psychologies and cultural patterns, to be sceptical of the wisdom of taking a populational view of culture: someone who thought that the fates of cultures could be reduced to the combined decisions of just a few very powerful people, or to structured networks of key actors, would also stand in opposition to these populational views. Or, to put the point another way, methodological individualists might sometimes oppose the use of populational models, and they might recommend functional decomposition instead.

Having distinguished these three forms of explanation in an idealized way, I will now make five observations about the relations between them. First, as indicated above, the distinction between functional decomposition and populational modelling is vague. Our examples of the functional decomposition of the parts of a car and the kinetic approach to gases constitute clear examples of the extremes of each approach, but there are explanatory approaches that lie in a borderline zone. One of these is the modelling and analysis of networks of individuals. Unlike the kinetic theory of gases, where it is essential that the numbers of molecules in a volume of gas are very large, network analysis often deals with small numbers of individuals, and the individuals in question may well have distinctive properties. For example, different individuals may have different levels of influence over others in the network: some may be largely isolated from the others, some may be connected to a great many different individuals, and these forms of interaction and connectedness can be subject to change over time. Thus, the analysis of a network has some features in common with functional analysis, but unlike the case of the car, where different parts have radically different functional roles and properties, the different individuals in these modelled networks are nonetheless conceived of (at least in some models) as generic points, which can vary merely with respect to their degrees of connectedness to other elements of the network.

Second, it should already be obvious that while, in practice, many explanatory models used by cultural evolutionists seek to explain the behaviours of interacting collections of individual people, there is no reason in principle why models cannot seek to understand the behaviours of interacting entities at different levels of organization. One could model the fate of a population of hunter-gatherer bands, or other supra-individual collectives. Turchin, for example, proposes a model for the formation of empires in which iterated interactions between rival communities of nomadic pastoralists and settled agriculturalists result in increasingly large, militarily powerful polities (Turchin 2009b).

Third, different explanatory techniques may be appropriate at different levels of organizational analysis (cf. Godfrey-Smith 2012). The operation of a car is best understood by means of functional decomposition: we need to understand how its different functional elements combine to produce motion in the object as a whole. Evidently this does not entail that functional decomposition is also the best way to understand the overall behaviour of large collections of cars in a road traffic system: here, a generic characterization of the typical car is likely to suffice. To turn to the human domain, an anthropologist studying interactions over time within small families might have little use of populational models, resorting instead to an analysis that takes note of distinctive features of different family members, and the relations those family members enter into. That would not rule out the application of populational models at the level of larger collectives of agents.

Fourth, while it is reasonable to characterize a large proportion of cultural evolutionary work as taking a 'kinetic', or 'populational', approach, I do not mean to imply that researchers interested in cultural evolution are never interested in looking at detailed patterns of interactions between individuals, or that they only ever focus on large-scale populational changes, or that the only explanatory models they make use of are of the populational variety. Primatologists interested in local cultural traditions in chimpanzees have made productive use of models that instead focus on the properties of small networks, where the individuals in those networks are given very specific characterizations.

As one might expect, observation of learning in these groups illuminates inequalities of power: Horner et al. (2010), for example, found that when members of their captive group of chimpanzees were faced with two different individuals from whom they might learn, they were more likely to copy the model who was older and had higher social rank. Hobaiter et al. (2014), whose work we encountered in Chapter 5, have moved the analysis of social networks into wild populations of chimpanzees. They have observed interactions between the Sonso chimpanzee community in Uganda in painstaking detail. Their study focused on chimpanzees' use of sponges—made either from chewed-up leaves or bits of moss—to soak up drinking water. Their observations of the chimpanzee group allowed them to generate a detailed representation of the network of learning events that had led to an entirely novel behaviour—namely the use of a sponge made from moss, rather than from the usual leaves—spreading through the group. They conjecture that they may have witnessed the very first use in the group of a moss sponge. (The Sonso chimpanzees have been studied for twenty years, but no investigator had ever seen moss used as a sponge before.) They also suggest it may have been significant that this first moss-sponging was achieved by an alpha male, and they note that it was observed by an adult dominant female. Again, attention to the details of interaction between individuals suggests the role of inequalities of power: the social rank of an innovator may affect the future prospects of that innovation in the population. In other words, Hobaiter et al.'s study further underlines the thought that—even for primates—the fate of a behaviour in a population may depend on idiosyncrasies of which individuals happen to encounter which others.

The fifth and final point worth making on the basis of these detailed studies of learning networks is that they have the potential to inform populational forms of modelling. Hobaiter et al.'s (2014) study provides evidence, on the basis of detailed observations of wild chimpanzees, regarding the nature of learning processes in these groups. Equipped with a better understanding of how learning works, we can then put together populational models whose assumptions are better confirmed empirically. It would be a mistake, in other words, to assume that if we are in the business of understanding the detailed and idiosyncratic structures of networks, we must oppose the use of populational models.

7.7 Revisiting the Reformation

Let me close by returning to Lewontin and the Reformation in Europe. Populational approaches have characterized many—but not quite all—forms of work in cultural evolutionary theory. One need not think that leaders of the various German states had total power to impose religious views on their subjects to be sceptical of the value of taking a fully 'kinetic' approach to cultural change. Rublack's (2004) history of the Reformation in Europe does indeed place stress on the ways in which individuals were exposed to, and influenced by, Lutheran and Catholic world views. But she also stresses the personal biographies of Charles V, Luther, and Calvin; the geographical and organizational particularities of Wittenberg and Geneva; the networks of influence fashioned by the trainees of Luther and Calvin; the ways in which the structure of a university enabled Luther to gain power for his ideas; and the ways in which Calvin instead relied on his *Compagnie des pasteurs*. This history, in other words, combines concern for the power of highly idiosyncratic individuals with concern for the power of complex organizations.

Populational approaches struggle to encompass both sets of phenomena. By contrast, techniques for understanding the behaviour of networks—techniques that remain in the minority in the corpus of cultural evolutionary work—can help us to understand better the implications of imbalances in power between individuals. When, as in some of

Turchin's analyses, idealized models do acknowledge power imbalances, they are typically addressed to the explanation of broad patterns that emerge across long swathes of history, as opposed to the detailed mechanisms of organizational power. To the extent that historians, or social anthropologists, are interested in understanding individual power and the power of complex organizations, it is not surprising that they have failed to see how the mainstream of cultural evolutionary theory might help them. Of course, that is not a defect of cultural evolution, for different problems require different tools. It does, however, neutralize any fears that evolutionary approaches to culture might overrun the traditional domains of social science.

8

Cultural Adaptationism

8.1 Evolution and Adaptationism

Many philosophers have been hard on the form of evolutionary psychology associated with Cosmides and Tooby's 'Santa Barbara school' (Tooby and Cosmides 1992). Several books and articles have criticized the approach for what, it has been claimed, are naïve assumptions about both evolutionary and psychological processes (e.g. Lloyd 1999, Dupré 2003, Buller 2005, Richardson 2007). Moreover, the most prominent theorists of cultural evolution have sometimes aimed to distance themselves from Cosmides and Tooby's approach, by claiming significant differences in the attitudes they take to cultural variation (e.g. Richerson and Boyd 2005: 10). Perhaps for that reason, philosophers have tended to embrace cultural evolutionary theory more warmly than evolutionary psychology. I shall argue in this chapter that we must be wary of imposing double standards in our assessment of these two evolutionary schools. There are, of course, very significant differences between them. But they also share much in common, including a commitment to various forms of adaptationist thinking.

Santa Barbara-school evolutionary psychology has often been attacked for its reliance on a form of methodological adaptationism. The question of how we should understand adaptationism has been treated in detail by me and by others, and I will not go into any detail about those debates here, except to say that methodological adaptationism is not a hypothesis about the forces affecting the organic world (Lewens 2009c). It is not the empirical claim that natural selection is the only important factor affecting evolving populations, nor is it the claim that natural selection inevitably produces well-designed traits. Methodological adaptationism of the kind that interests me in this chapter instead offers a recipe for biological enquiry. It recommends that consideration of the circumstances in which our species evolved is likely to provide us with various heuristic insights when it comes to formulating hypotheses about how our minds and bodies work right now. In the psychological context, it is the view that reflection on the demands of ancestral human environments gives us an epistemic leg-up when we reflect on modern psychology. In this chapter I begin by making a purely descriptive claim: this same methodological adaptationism also characterizes much of the work done by cultural evolutionary theorists. I then move to ask what, if anything, is wrong with this form of adaptationism?

8.2 Boyd and Richerson Meet Cosmides and Tooby

It might tempting to say that cultural evolutionists like Richerson and Boyd stress the reality of cultural diversity, while Cosmides and Tooby instead stress the psychological unity of mankind. This effort to draw a contrast between the two schools obscures important commonalities. First, we have already seen that Richerson and Boyd sometimes suggest— whether for the purposes of idealization in their formal models, or for ease of presentation in their informal writings—that the cognitive dispositions that bias how we learn from each other are broadly shared across our species. Second, we should be clear that Cosmides and Tooby are not in the business of denying that cultural variation is real. Instead, they stress, rather in the manner of Dan Sperber, the compatibility between the existence of a collection of more or less universal and unchanging cognitive adaptations and the existence of cultural diversity and cultural change as those constant adaptations react in different ways because they are exposed to different cultural settings. So, to rehearse a well-worn but helpful example, Cosmides and Tooby suggest that a preference for sweet foods has remained intact since the Pleistocene, but the expression of that preference has changed significantly as fast-food restaurants have become ubiquitous (Cosmides and Tooby 1997). The modern epidemic of obesity is thereby understood as the result of unchanging adaptations encountering an altered environment.

Sperber puts Cosmides and Tooby's point like this, in a manner that stresses how the mere recognition of a universal, evolutionarily endowed mental potential does not preclude rich, local, and contingent cultural variation:

> Generation after generation, humans are born with essentially the same mental potential. They realize this potential in very diverse ways. This is due to the different environments, and in particular to the different cultural environments, into which they are born. However, from day one an individual's psychology is enriched and made more specific by cultural inputs. Each individual quickly becomes one of the many loci among which is distributed the pool of cultural representations inhabiting the population. (1996: 115)

So Richerson and Boyd are by no means committed to denying the existence of more or less universal elements of our psychology, which owe their ubiquity to evolutionary processes; meanwhile, Cosmides and Tooby (and Sperber) are equally uncommitted to any sort of implausible denial of cultural variation.

If these are the broad-grained similarities, what distinguishes Boyd and Richerson from Cosmides and Tooby? Most obviously, and most importantly, the work of cultural evolutionists has tended to focus on demonstrating the existence of, and evolutionary rationale for, various learning biases: in other words, they have focused on alleged cognitive tendencies—such as prestige bias or conformist bias—that are thought to affect which individuals we learn from.

Cosmides and Tooby, by contrast, have tended to focus more on describing alleged tendencies to process some kinds of information more readily than others. In other words, they have focused their analysis on content biases. They famously argue, for example, that we are better at evaluating claims about social contracts, and whether they are being violated, than we are at evaluating other claims with the same syntactic structure, but different semantic content. Roughly speaking, then, one evolutionary camp emphasizes whom we acquire our representations from, while the other emphasizes what sort of representational content we find easiest to handle.

8.3 Adaptive Lag

In their own efforts to distance themselves from Cosmides and Tooby, Richerson and Boyd stress two themes. First, they put considerable emphasis on the creative role of social learning. As we have already seen with the case of lactose tolerance, cultural changes are not merely the causal consequences of genetic adaptations reacting to altered environmental settings; they are also the causal progenitors of new genetic adaptations. Richerson and Boyd credit social learning with a form of adaptive creativity in its own right. They argue, in part on the basis of the models we encountered in Chapter 6, that when individuals make use of a suitable combination of individual and social learning, the information they acquire means they are better able to act appropriately in their surroundings. As they say, it would be 'nuts' (2005: 46) to suggest that natural selection acting on genetic variation has shaped our species so that we are able to develop, without significant effort, the sorts of complex skills involved in hunting and gathering. Instead, knowledge of where to find safe sources of water, how to track game, how to fashion weapons, how to build adequate shelters, and so forth are all hard-won skills that are indeed shaped over generations, but which are created and maintained by various forms of effortful learning and the vigilant practice of tradition.

Second, Richerson and Boyd point out that, on their view, maladaptive behaviour should not always be written off as the 'misfiring' of a once adaptive mechanism in the face of an altered environment to which it is no longer well suited. Cosmides and Tooby appeal to this sort of 'adaptive lag' to explain the modern epidemic of obesity. But consider the case of prestige bias, which Richerson and Boyd characterize as a disposition to imitate the successful. There is some empirical evidence for the reality of prestige bias. Henrich and Broesch (2011) have argued, on the basis of fieldwork in Fiji, that an individual's perceived success in a single domain of activity (for example, yam cultivation), predicts whether that individual will be asked for advice in other domains (for example, fishing). In other words, they claim that individuals are accorded a broad form of prestige, which affects their likelihood of being imitated. Richerson and Boyd argue that if individuals copy techniques from those who are in prestigious positions, then this increases the chances that they will copy techniques that are, in fact, beneficial. As they put it, 'Determining *who* is a success is much easier than determining *how* to be a success' (2005: 124).

The adaptive value of prestige bias relies on the supposition that those individuals who are able to get themselves into prestigious positions have a better than average tendency to make use of fitness-enhancing techniques. This heuristic will not be failsafe: after all, not every technique a prestigious individual uses will also augment fitness, and some individuals may be accorded prestige without good cause. But the question that settles whether natural selection might promote prestige bias is not whether prestige bias will sometimes lead to the copying of maladaptive techniques; the question, rather, is whether individuals who learn from the prestigious will tend to be fitter on average than individuals who either do not learn at all, or who are equally likely to learn from any member of the population regardless of their social status. In summary, Richerson and Boyd deny that it is only now—in a novel cultural context, where meritless celebrities are unavoidable—that prestige bias is likely to enable the spread of maladaptive traits. It always had this negative effect, but these costs were outweighed by its benefits (Richerson and Boyd 2005: 166). Adaptive lag is not always the explanation for maladaptive behaviour.

These differences may reflect varied emphasis between evolutionary schools, rather than profound rifts. After all, Sperber generally allies his approach to the evolved mind with that of Cosmides and Tooby, but a stress on the importance of content biases does not (and should not) lead him to deny that important bodies of information are built up over time by learning from others, and he openly embraces the idea that cultural changes can have knock-on effects on genetic evolution. Whatever our final reckoning might be regarding these apparent differences, we should not be blind to deep similarities between the three schools of Boyd and Richerson, Sperber, and Santa Barbara, to which I now turn.

8.4 Cultural Evolution's Adaptationism

As we noted in Chapter 1, all three schools espouse a form of adaptationism. More specifically, all three schools endorse an image whereby natural selection acting on genetic variation in the Pleistocene has built a series of cognitive adaptations that are characteristic of our species and have not changed much since that time. Moreover, they all recommend that reflection on the demands of these ancestral environments can help us when we seek to understand how our minds work right now.

The heuristic value of this form of methodological adaptationism is most clearly expressed by Cosmides, Tooby, and Barkow:

By understanding the selection processes that our hominid ancestors faced—by understanding what kind of adaptive problems they had to solve—one should be able to gain some insight into the design of the information-processing mechanisms that evolved to solve these problems. (Cosmides et al. 1992: 9)

Sperber's adaptationism is also well known. He endorses Cosmides and Tooby's general evolutionary approach to the mind (Sperber 1996: 113). He also shares their further conviction that reflection on our evolutionary past can assist us in determining the organization of human minds, and more specifically that it can help us to understand the sorts of mental modules they contain:

I have been convinced by Leda Cosmides and John Tooby...that we know enough about evolution, on the one hand, and cognition, on the other, to come up with well-motivated (though, of course, tentative) assumptions as to when to expect modularity, what properties to expect of modules, and even what modules to expect. (Sperber 1996: 124)

Similarly, when Henrich and Boyd (1998: 216) write that 'there has been little effort to explain the existence of cultural variation between groups in terms that are consistent with the assumption that the psychological mechanisms that create and maintain such variation are evolved adaptations', it is clear that they understand adaptations to be the products of natural selection acting on genetic variation. As we saw back in Chapter 1, they share Cosmides and Tooby's conviction that evolutionary processes have a tendency to construct modular cognitive adaptations in response to environmental problems. Recall, in particular, their contention that:

Cultural transmission mechanisms represent a kind of special purpose adaptation constructed to selectively acquire information and behavior by observing other humans and inferring the mental states that give rise to their behavior. (Henrich and Boyd 1998: 217)

McElreath et al. express a similar sentiment when they tell us that 'because the psychological mechanisms that make social learning possible are partly products of natural selection, evolutionary models are necessary to fully understand their design' (2005: 484). Richerson and Boyd take the view that our learning biases were favoured among ancestral humans because they worked well as solutions to problems posed by the environment: 'Culture', they say, 'was originally an adaptation to Pleistocene climate chaos' (2005: 146).

At the same time as arguing that the cognitive traits that underpin culture have been shaped by the demands of the Pleistocene, Richerson and Boyd express optimism about the ability of human cognition to perform well under modern conditions of operation: 'Human culture as an adaptive system evolved in response to Pleistocene environments but has subsequently upped anchor and sailed rather well on uncharted waters' (2005: 147). In saying this, they are not claiming that learning dispositions shaped in the Pleistocene have changed under the influence of cultural shifts, in ways that have maintained their adaptive fit with their altered modern surroundings. Instead they are claiming that even if they have not changed much, learning dispositions shaped during the Pleistocene often turn out to function rather well in new cultural surroundings.

In their discussion of the emotions that regulate social behaviour, for example, they endorse Cosmides and Tooby's own suggestion that there has probably not been enough time since the Pleistocene for natural selection acting on genetic variation to have altered our 'social instincts'. The result is that these instincts have not changed much over thousands of years:

The increase in the size and complexity of human societies has probably not been accompanied by significant changes in our social instincts. While natural selection can sometimes lead to substantial genetic change in a few thousand years, most biologists think that important changes in complex characters take much longer to assemble. *Our innate social psychology is probably that bequeathed to us by our Pleistocene ancestors.* (Richerson and Boyd 2005: 230, emphasis added)

There is, then, an important sense in which Richerson and Boyd agree with Cosmides and Tooby's (1997) notorious assertion that 'our modern skulls house Stone-Age minds'.

I have argued so far that Richerson and Boyd (and Sperber, too), all share with Cosmides and Tooby a broad methodological adaptationism, which recommends that we should reflect on the demands of past environments in order to understand better our cognitive make-up today. This mode of reflection is partly justified through the thought that our cognitive adaptations will not have changed much since the Pleistocene. I now turn to the issue of whether there is anything wrong with this form of adaptationism.

8.5 Externalism

Let us begin this assessment by quickly setting one issue aside. Adaptationism is sometimes understood as a predominantly 'externalist' mode of explanation (e.g. Godfrey-Smith 1996). Industrial pollutants made the trees around Birmingham darker, and moths became darker to match them; more generally, environments change, and species eventually match those changes, or so the story goes. The broad idea here is that the adaptationist is committed to a form of explanation whereby organic lineages change as the result of external environmental forces.

It is debatable whether adaptationism has ever really been committed to this externalist stance (Lewens 2004). Adaptationists have long stressed notions of frequency-dependent selection, for example, whereby the successes of individual organisms bring about altered selection pressures. The cultural evolutionary project makes especially vivid the error of thinking that adaptationism entails externalism. We have already seen in this book that adaptationists do not merely ask how organisms will change to fit their environments. They also ask how organisms might change their environments to their own benefit. Moreover, adaptationist reasoning is increasingly applied to processes of development and inheritance in their own right. Many of Richerson and Boyd's models aim to explain, via appeals to selection, why it should be possible for organisms to learn from each other at all: in other words, they ask after the ways in which developmental processes might change under the influence of selection. Kim Sterelny's thinking about cultural inheritance has for many years been characterized by an adaptationist stance on the sorts of developmental processes that facilitate the evolution of novelty (e.g. 2001). Considerable bodies of work in more general theoretical biology have examined the manner in which selective processes affect each other at different levels of biological organization, and the ways in which selection may affect the capacity of developing systems to change over time in cumulative ways (Buss 1987, Maynard Smith and Szathmary 1995, Jablonka 2001, Jablonka and Lamb 2005).

So adaptationism—both in its general form, and in its specific applications to cultural evolution—should not be dismissed on the grounds that it inevitably distracts our attention away from important processes of development, and it should not be dismissed on the grounds that it assumes that inheritance processes themselves are fixed (Lewens 2009c). Adaptationism is not generally prey to accusations of undue externalism, and cultural evolutionary work illustrates this well.

8.6 Epistemic Circles

Cultural evolutionary work is vulnerable to two rather different criticisms that have long been associated with adaptationism. The first goes back at least as far as Gould and Lewontin (1979), and has been pressed by Philip Kitcher (1985) in the context of sociobiology, and by David Buller (2005) in the context of evolutionary psychology. All have been concerned that evolutionary work is sometimes content to loop around overly tight epistemic circles. These circles give spurious confirmation to the existence of traits whose grounding in experiment is only tentative, on the basis that rather sketchy evolutionary models predict that these same traits would 'make sense' from the perspective of natural selection.

Gould and Lewontin gave a nice example of the dangers of this sort of approach back in 1979, when they criticized a piece of David Barash's work on birds:

Barash mounted a stuffed male near the nests of two pairs of bluebirds while the male was out foraging. He did this at the same nests on three occasions at ten-day intervals: the first before eggs were laid, the last two afterwards. He then counted aggressive approaches of the returning male toward the model and the female. At time one, aggression was high toward the model and lower toward females but substantial in both nests. Aggression toward the model declined steadily for times two and three and plummeted to near zero toward females. Barash reasoned that this made evolutionary sense, since males would be more sensitive to intruders before eggs were laid than afterward (when they can have some confidence that their genes are inside). Having devised this plausible story, he considered his work as completed. (Gould and Lewontin 1979)

Gould and Lewontin then raise an obvious problem for Barash's interpretation of his results: it might be that aggression declines because the certainty of paternity increases; but it might also be that aggression declines because the male bluebird realizes that the apparent interloper is just a stuffed fake.

While Gould and Lewontin's criticism may cause trouble for Barash's particular piece of work, it has no general bite against biologists who are interested in formulating and testing claims about adaptation: it merely reminds us that some pieces of work in sociobiology were too hasty in assuming that their hypotheses were well confirmed. The antidote to this worry is not to abandon the adaptationist programme; it is to ensure that evidence is gathered in a more rigorous way.

Gould and Lewontin were concerned about the illusory form of confirmation that arises when loose evolutionary reflection comes into alignment with loose experimental test. One might assume, partly because of their use of quantitative evolutionary modelling, partly because of a more rigorous approach to experiment, that these sorts of worries never arise when modern cultural evolutionists formulate and test hypotheses about cognitive adaptations. But our detailed discussion of conformist bias in Chapter 6 suggests that the use of mathematical models, rather than more intuitive forms of speculation, does not suffice to neutralize Gould and Lewontin's concerns. Here, too, the strong initial impression one gets from this work is that rigorous mathematical modelling tells us that conformist bias is the sort of trait that would have been beneficial to our ancestors under a very broad range of plausible environmental conditions; meanwhile, direct empirical work suggests that conformist bias is a real feature of human psychology. So, once again we appear to have an evolutionary hypothesis that prompts, in a heuristically valuable way, the investigation of psychological phenomena. Those phenomena are then found to fit with (and thereby confirm) the evolutionary hypothesis. But closer inspection reveals that the evolutionary models themselves derive their results from questionable simplifications, and empirical confirmation for the very specific form of conformity required by those models remains mixed. It turns out that some prominent cultural evolutionary work is still vulnerable to the very same concerns about adaptationism that motivated Gould and Lewontin's sceptical treatment of Barash.

8.7 Adaptive Thinking

The second long-standing problem laid at the door of a certain kind of adaptationism concerns the extent to which we should expect reflection on past environmental demands to illuminate the workings of our minds today. The empirical search by cultural evolutionists for prestige bias, conformist bias, and so forth is prompted by intuitions, driven by considerations of potential adaptive benefit to our ancestors, that such biases are likely to be widespread in modern populations. As Richerson and Boyd put it, 'Darwinian analysis reveals a mass of largely unexplored questions surrounding the psychology of cultural transmission and the biases that affect what we learn from others' (2005: 123). In other words, cultural evolutionists frequently couple reflection—enhanced by theoretical modelling—about the sorts of cognitive adaptations that would probably have been favoured in the Pleistocene to empirical work on extant populations that subsequently aims to confirm the reality of these putative adaptations.

Concerns about this form of 'adaptive thinking' are well known from discussions of methodological adaptationism in evolutionary psychology (e.g. Buller 2005), and it would be tedious to repeat these old arguments in great detail here. The questions at stake are not whether our minds have evolved, whether natural selection is the most important evolutionary force acting on our minds, or whether our minds are likely to be collections of modular adaptations favoured according to their contributions to fitness in past environments. Even if the answer to all these questions turns out to be 'yes', we still can ask whether knowledge of the environment in which we evolved is likely to offer much help when we try to understand the mechanisms and dispositions of modern minds.

The identification of a well-confirmed set of ancestral adaptive problems, and of a plausible set of solutions that a lineage might throw up in response to those problems, presupposes that we know a significant amount about Pleistocene climatic conditions, botanical and zoological challenges, and the social demands imposed by navigating interactions with others. It also presupposes that we understand how ancestral human minds were organized, how they developed, and how ancestral human practices were acquired and executed (Griffiths 1996, Sterelny and Griffiths 1999, Lewens 2002b). Without these latter forms of knowledge we cannot put plausible constraints on how ancestral problems might have been solved, for the pathways that an evolving population will take depend, among other things, on how developing brains respond to genetic mutations, on how mutations affect the overall systemic organization of the developing organism, and on how broad environmental 'problems' trade off against each other when they come into conflict. It is difficult, in other words, to identify a set of ancestral problems with enough content and stability to offer the researcher much heuristic insight. The result is that claims about the directions in which we should expect cognitive evolution to have been directed by natural selection sometimes have an overly speculative character.

Jerry Fodor, for example, suggested long ago that an increasingly integrated, non-modular cognitive architecture would most likely have been favoured by evolution, on the grounds that if more diverse forms of information are brought together, inferences will be more reliable:

Cognitive evolution would thus have been in the direction of gradually freeing certain sorts of problem-solving systems from the constraints under which input analysers labour—hence of producing, as a relatively late achievement, the comparatively domain-free inferential capacities which apparently mediate the higher flights of cognition. (Fodor 1983: 43)

Dan Sperber responds with his own speculations, which run in the opposite direction. He concedes that demodularization might be advantageous in principle. But he insists we should not have too simplistic a view of evolutionary processes. Suppose that our early ancestors did indeed have several distinct, specialized cognitive modules. Sperber suggests that this early state will tend to give rise to a form of 'lock-in', which makes demodularization unlikely in practice. A root-and-branch reorganization of cognitive architecture may not be feasible, given the need for evolving systems to improve by gradual changes to existing set-ups. So, while he concedes that 'it might be advantageous to trade a few domain-specific inferential modules for an advanced all-purpose macro-intelligence, if there is such a thing', he goes on to caution that:

evolution does not offer such starkly contrasting choices. The available alternatives at any one time are all small departures from the existing state... A demodularization process is implausible for this very reason. (Sperber 1996: 126)

Hence his conclusion: 'evolved cognitive modules are likely to be answers to specific, usually environmental, problems' (1996: 127).

I want to offer two reasons for scepticism about how far we can get with this very high-level form of speculation about how evolution is likely to answer environmental problems. First, even when intuitions of efficient performance might lead us to expect that specialized traits will arise in response to specific environmental problems, there are many examples where, as a matter of fact, a specialized solution is not what ends up evolving. Our ancestors faced problems related to throwing spears, gathering fruit, and communicating with each other using gestures. The result of these very different problems is that they are all solved using the human hand: we have not acquired different appendages to confront each problem (Heyes 2012c).

Sperber is right, then, to point out that the degree to which natural selection will modify an ancient system will depend not just on the ideal solutions to the problems faced by a lineage, but also on the degree of constraint imposed on the system as a whole. But the question of the extent to which these constraints affect the evolving brain—a question which concerns the plasticity of the brain, its capacity for reorganization, and the ways in which various inherited developmental resources affect its ontogeny—is itself up for grabs.

Second, the 'adaptive thinking' we are considering here asks us to predict what sorts of cognitive mechanisms natural selection is likely to build in response to environmental problems. The intuition behind this heuristic is sometimes illustrated by appealing to our understanding of designed artefacts. The behavioural ecologists Krebs and Davies suggest that:

Visitors from another planet would find it easier to discover how an artificial object, such as a car, works if they first knew what it was for. In the same way, physiologists are better able to analyse the mechanisms underlying behaviour once they appreciate the selective pressures which have influenced its function. (Krebs and Davies 1997: 15)

If we understand this recommendation in a mild way then it is unobjectionable: without some empirically grounded hypothesis regarding the overall tasks a structure is able to accomplish, it is hard to see how we might generate conjectures for its internal working. But experimental work suggests that even when environmental problems are stable and simple, it can still be very difficult to predict the sort of mechanisms that natural selection will favour, in advance of their construction.

A vivid example comes from the domain of evolutionary electronics, where researchers allow natural selection to modify lineages of electrical circuits, selecting the ones that perform some pre-specified task best (Lewens 2013). Adrian Thompson's group from the University of Sussex used something called a field programmable gate array (FPGA)—essentially a reconfigurable chip—whose organization was determined by a genetic algorithm. They tested a series of slightly modified configurations to the chip, preserving the circuits that were best able to discriminate between two audio tones. Although the FPGA had a 64 × 64 matrix configuration, the programmed section of it used only a 10 × 10 area of one corner of the chip. After thousands of iterations, a version of the chip that performed the task more or less perfectly was bred. The exact details of the investigation do not matter here, but the final piece of hardware had some extraordinary properties. Most striking for our purposes is the fact that some sections of the best-performing evolved chip could not be altered without loss of function, even though they were not electrically connected (in the usual way, at least) to the output of the chip:

> [These cells] cannot be clamped [i.e. returned to their default 'blank' state] without degrading performance, *even though there is no connected path by which they could influence the output* [. . .] They must be influencing the rest of the circuit by some means other than the normal cell-to-cell wires: this probably takes the form of a very localised interaction with immediately neighbouring components. Possible mechanisms include interaction through the power-supply wiring, or electromagnetic coupling. (Thompson 1997, emphasis in original)

It would have been exceptionally difficult for researchers to predict the organization of the final evolved chip merely on the basis of their knowledge of the task they had set for it. Indeed, these researchers were very unsure about how the best-performing chip worked even when they could examine it directly. Unconstrained pathways of evolutionary design are free to take advantage of whatever fitness-enhancing effects may present themselves in a lineage. This should give us significant doubts about the ability of observers to predict how valuable effects will be achieved in evolved systems in advance, merely on the basis of knowledge of the problems posed.

In the case of Thompson's FPGA, the researchers themselves ensured a single, stable selection regime for their lineage of circuits, which consistently aimed at the solution of just one 'adaptive problem'. The difficulties researchers on human evolution have in predicting our cognitive organization are further compounded by the fact that organic lineages face a series of conflicting, and fluctuating, 'adaptive problems'. These worries do not trade on the idea that natural selection may be unimportant compared with other evolutionary processes. They would obtain even if the fittest variant always prevailed, and in Thompson's study experimenters ensured this was the case. Instead, our worries turn on the difficulties of using knowledge of past 'problems' to predict current 'solutions'.

I have argued in this section that broad speculation based on very general thoughts about past adaptive problems is likely to have little heuristic pay-off. This does not mean that adaptive thinking is invariably worthless. If we can construct detailed models of human minds and human social action, and if we can couple these models with plausible developmental hypotheses regarding the sorts of cognitive variations that might be available in evolving populations, then heuristic insight regarding potential responses to ancestral problems is more likely. But that just means that our best guesses for how our minds and practices used to work in past environments will need to be strongly informed by the disciplines that tell us how they do work right now. The cultural evolutionist's adaptive thinking should be highly deferential to work in developmental psychology, neuroscience, ethnography, and so forth. These disciplines will not be eclipsed by cultural evolutionary work: on the contrary, if cultural evolutionary work is to make progress, its practitioners will need to steep themselves in these more traditional approaches.

8.8 Stone-Age Minds

When Cosmides and Tooby defend the utility of reflecting on past selective environments for shedding light on modern minds, they appeal to the thought that there has not been enough time since the Pleistocene for significant adaptive evolution to have occurred. The result, they say, is that our minds now are more or less the same as they were back then. As we have seen, Richerson and Boyd sometimes suggest that we have inherited a set of 'social instincts' from our Pleistocene ancestors, which have also gone unchanged over this relatively brief period of evolutionary time. To repeat a quotation we saw earlier, they point out that:

While natural selection can sometimes lead to substantial genetic change in a few thousand years, most biologists think that important changes in complex characters take much longer to assemble. Our innate social psychology is probably that bequeathed to us by our Pleistocene ancestors. (Richerson and Boyd 2005: 230)

The issue of how quickly genetic evolution can build complex adaptations is a contentious one, but even if we agree with Richerson and Boyd that this process is most likely too slow for selection to have changed our cognitive adaptations much since the Pleistocene, we cannot rule out the thought that these adaptations will have changed as the environments that affect their development change (cf. Hrdy 2009: 290–294). If, for example, the development of cognitive adaptations turns out to be sensitive to changeable social and cultural circumstances, then we have reason to think these adaptations can change—perhaps significantly—as social and cultural environments change.

As we saw earlier, Sperber is well attuned to the ways in which alternative developmental environments affect adult cognitive make-up. He hints that differences in cultural environments do not merely affect which representations populate the minds of different people, but that these different environments affect the very mental processes that determine how such representations are handled. As we saw in Chapter 4, Cecilia Heyes (2012a) presents empirical evidence suggesting important roles for learning in the ontogeny of, among other things, the capacity of humans to imitate each other. Heyes conjectures, in other words, that functional features of adaptive learning capacities are shaped by their social contexts of development. If this is the case right now, we cannot rule out the thought that cultural contexts have informed the development of learning biases across long swathes of human evolutionary time. That, in turn, opens up the possibility that human cognitive adaptations have changed as the cultural circumstances of their development have changed.

Heyes has suggested that the adaptive development of the capacity to imitate may be enhanced by, among other things, the use of mirrors and the location of the growing individual in an appropriate social environment. Evidently mirrors have not always existed, and social circumstances have not been stable. So we must not ignore the possibility that our cognitive adaptations differ from Pleistocene adaptations because the social environments in which they are learned—and in which they develop in other respects—have changed markedly since that time. Human cognition presents the evolutionist with a moving target.

8.9 The Cultural Development of Cultural Adaptation

The impression that Richerson and Boyd are committed to an image of the mind as a collection of innately specified adaptations for social interaction may be an artefact of some casual features of their informal writings. They sometimes mention 'social instincts', and they suggest these instincts are 'innate', but it is doubtful that these claims—which amount to hypotheses about how these traits develop—are central to their true interest, which is to secure a plausible functional rationale for prosocial emotions. Elsewhere, when discussing learning biases, they clearly acknowledge that learning strategies develop in ways that are contingent on social and cultural surroundings. Peter Richerson, for example, is one of many co-authors in a paper that warns:

> We do not imagine that social learning strategies, which themselves can be learned, are invariant human universals. The strength of conformity, in particular, likely varies cross-culturally and situationally. (McElreath et al. 2005: 506).

Prefiguring Henrich et al.'s important work on WEIRD (Western, educated, industrial, rich, democratic) societies, they go on to caution, quite rightly, that we cannot assume that what we learn about the social learning strategies of American students can be generalized to all people who learn from others:

> Students in Western societies are repeatedly admonished to 'think for themselves'. It is also important to notice that students, the favorite subjects of psychologists and economists alike, are an odd population to study to understand how people learn. Students in university are trained to learn in particular ways that are unlikely to be representative of most adults. Constructing theories of human nature based on student data is always hazardous, but particularly so in this case. (Ibid.)

All of these caveats are sensible. How we learn can be affected by training, training differs from culture to culture, and so the ways in which we learn differ across the globe, and across time. Similar worries have been stressed regularly by social anthropologists: Descola, for example, noted back in 2005 (although his work was not translated into English until 2013) that supposedly universal 'cognitive schemas' have been inferred 'on the basis of experiments conducted almost exclusively in Western industrialized societies' (2013: 102). To the extent that we take these caveats seriously, we accentuate our reasons for thinking that cognitive adaptations are unstable over time, precisely because the developmental processes that construct them in each generation are subject to significant perturbations. In other words, we further weaken the degree to which reflection on the demands of ancient environments, where universities did not exist, and where societies were very different, will inform us about learning in the modern world.

8.10 Cultural Phylogeny

The suggestion that we have inherited a set of cognitive adaptations, fashioned in the Pleistocene, which are largely impervious to modification by altered circumstances of learning and enculturation, is an unwise one

(Mace and Jordan 2011). What impact does this have on the likely shape of a successful evolutionary approach to culture? The project of tracing the pathways by which our lineage has come to its modern capacities through the modification of their ancient progenitors remains intact. This natural historical endeavour, pursued for several years now by the likes of Sterelny, Jablonka and Lamb, and others, and explored in detail in Chapter 9 of this book, argues that reflection on the past can help illuminate the present, while denying that our minds are mere replicas of those of our ancestors. Similarly, Darwin's own exploration of the emergence of the moral sense in humans told a plausible history by which early capacities for beneficent behaviour were transformed into a system of public morality, while acknowledging that our moral psychology was still in a process of change.

These historical projects are important, and the tools of cultural phylogeny—which we have only mentioned briefly, and whose detailed investigation raises questions that lie outside the scope of this book—can assist them. Just as close resemblance with respect to functionless sequences of DNA is well explained by common ancestry of species, so resemblance of texts with respect to typographical errors or resemblance of rugs with respect to arbitrary decorative features are also strongly indicative of a common historical origin.

We have inherited from Darwin an image of evolutionary change as following a pattern of tree-like divergence. Cultural change, by contrast, is not so well behaved. A complex object like a car does not evolve independently of other technologies. It constantly lends to, and borrows from, other technical lineages. Cars have incorporated elements of the microcomputer, the laser, the horse-drawn carriage, and so forth. While the likes of Gould (1988) have consequently been sceptical of the value of phylogenetic tools in the cultural domain, cultural evolutionists themselves point out the degree to which much of the biological world also misbehaves when measured against the reassuring shape of Darwin's trees. Bacteria do not form genealogically isolated lineages, hybridization is rife among plants, and there may also be considerable borrowing of elements of the genome between apparently isolated mammalian species. Horizontal gene transfer, in other words, is ubiquitous. This might show simply that phylogenetic modes of inference are doubly imperilled: they do not work for much of the biological world either. But cultural evolutionists are heartened by developments within biology itself. Suppose a portion of genetic sequence is frequently transferred horizontally across several distinct species. This will have the result that a genealogical tree for that sequence will not coincide with proposed trees for the species in question that draw on broader collections of genetic sequence data, or morphological data. Newer biological techniques aim to reconstruct partially reticulated trees by proposing 'reconciliations' of the conflicting trees that traditional methods often propose for species and genes (Gray et al. 2007).

Various phylogenetic tools have been fruitfully applied to diverse samples of cultural items—everything from texts (Barbrook et al. 1998) to textiles (Matthews et al. 2011)—to reconstruct their own likely patterns of borrowing. That does not mean, of course, that a phylogenetic analysis alone will tell us anything about what different versions of Chaucer's *Canterbury Tales* meant to the people who read them, but it can help us with the pure historical project of understanding which of a variety of texts most likely served as the originals from which others were copied (Barbrook et al. 1998). Such a phylogeny also leaves open many other questions that would interest investigators from social-scientific backgrounds: why have Chaucer's texts been maintained at all, and what might explain choices about which earlier versions to copy from? That said, once such phylogenetic tools have been used to establish plausible histories, these histories can constrain further functional hypotheses (Mesoudi 2011, O'Brien et al. 2012). That role is also reflected in mainstream biology. Why do the various species of bear have comparatively low birth weights? Perhaps the answer is that low birth weight is a concomitant of the physiological requirements of hibernation. But Griffiths (1996), following McKitrick (1993), shows how phylogenetic analysis of bears and their relatives establishes that low birth weight emerged prior to hibernation, and that it exists in branches of the phylogenetic tree in which hibernation never evolved. Phylogeny, therefore, undermines the claim that low birth weight is to be explained as a functional trade-off linked to hibernation.

Having said all this, one must not exaggerate how much can be learned about function even from a well-confirmed cultural phylogeny (cf. Borgerhoff Mulder 2001). As Darwin remarked in the *Origin*:

The sutures in the skulls of young mammals have been advanced as a beautiful adaptation for aiding parturition, and no doubt they facilitate, or may be indispensable for this act; but as sutures occur in the skulls of young birds and reptiles, which have only to escape from a broken egg, we may infer that this structure has arisen from the laws of growth, and has been taken advantage of in the parturition of the higher animals. (Darwin 1859: 197)

Darwin reminds us that even if we could show, on the basis of an accurate phylogeny, that cranial sutures did not appear in mammals in order to facilitate parturition, this would leave open the thoughts that mammalian sutures do indeed facilitate parturition, that they have been maintained by selection because of this role, and consequently that they can properly be thought of as adaptations for parturition (cf. Sober 2009, Lewens 2015). A claim about function is undermined here only to the extent that we initially justified it on the basis of a supposed correlation across mammals and other taxa between (for example) live birth and cranial sutures. If, instead, we simply observe directly that the heads of mammals do indeed squash up as they pass through the birth canal, and that, when this is impossible, the risks of problems in childbirth are far greater, then our source of evidence for function lies elsewhere, and phylogeny does far less to damage our functional hypothesis.

To turn back to the cultural domain, there has been much anthropological speculation about the functions of various practices by which money and other valuable resources are transferred at marriage. If we have data showing a strong correlation across diverse societies between polygyny and bridewealth (i.e. the transfer of property from a groom, or from his family, to the family of the bride), and a strong correlation between monogamy and dowry (where property is instead transferred to the bride, from her own family), then this supports the hypothesis that bridewealth is a functional response to polygyny (on the grounds that one's sons can purchase many marriage partners), while dowry is a functional response to monogamy (on the grounds that one's daughter can secure the best possible husband). These correlational data are brought into question, though, by 'Galton's problem': if phylogenetic analysis tells us that the societies where polygyny and bridewealth are found together have most likely inherited these conjoined practices from a shared historical ancestor, then our correlational data no longer suggest that bridewealth has repeatedly arisen in response to polygyny. That, in turn, means our correlational data no longer offer such strong support to a hypothesis about the function of bridewealth (Fortunato and Mace 2009). The tools of cultural phylogeny can help to combat, in a variety of ways, the epistemic problems that beset simple adaptationist reasoning about supposed evolutionary responses to putative ancestral scenarios (Griffiths 1996). But while cultural phylogenies can helpfully inform claims about function, we should not suppose that they are invariably decisive in such debates.

8.11 Towards an Eclectic Synthesis

We have seen reason to reject the idea that reflection on the general demands faced by our ancestors in the Pleistocene will offer much heuristic insight when we come to investigate how our minds function right now. Our discussion of cultural adaptationism and cultural phylogenetics reinforces the case, tentatively made in Chapter 2, for an eclectic synthesis in cultural evolution. I have not argued that adaptive thinking is without value. I have argued that the application of these forms of thought must be strongly deferential to work that studies the detail of human development and social interaction. Any plausible story about the shaping of human capacities for thought, action, innovation, and persuasion must be strongly constrained by the best work not only from natural scientific disciplines such as neuroscience and psychology, but also from social-scientific disciplines such as social anthropology, and even from disciplines that are rarely classified among the sciences at all, such as history. These latter disciplines tell us about how humans interact—and hence how they develop, cohere, and are transformed—outside the laboratory and in the wilds of day-to-day life.

9

Eclectic Evolution
The Case of the Emotions

9.1 Emotion in Cultural Evolution

Chapter 8 reinforced the case for an eclectic approach to cultural evolution. This final chapter uses the example of the emotions to illustrate what such an eclectic approach might look like. Surprisingly, cultural evolutionary theorists have had comparatively little to say about the emotions. There is, of course, a significant body of evolutionary psychological work in this area, including Ekman's well-known research on the universality of the so-called basic emotions (e.g. Ekman 1992), and work on love understood as a commitment mechanism (Frank 1988). But this work usually takes the view, more or less overtly, that the emotions in question are to be understood as evolving under the control of genetic variation. There is also a tradition of historical research that documents changes in emotional categories—it addresses, for example, the emergence of the term 'emotion' itself and the diminution of older terms such as 'passion' (e.g. Dixon 2003)—but such work is usually hesitant when it comes to claiming changes in the states those categories aim to classify.

Although comparatively little cultural evolutionary work has addressed the emotions, there is some. Richerson and Boyd's 'tribal social instincts hypothesis', for example, aims to explain why individuals should feel deep emotional commitments to the groups of which they are a part, even when these commitments are in conflict with other emotional ties that may be directed to self or kin (Richerson and Boyd 2005, inter alia). Cultural processes play an essential role in the explanation they offer. In brief terms, their story begins with natural selection promoting conformist bias, by virtue of its ability to allow individuals to learn useful information from their fellows. Once a disposition to conform is widespread, the power of natural selection to act on cultural groups becomes enhanced. For if humans have a tendency to conform to the dominant group trait, migration away from or into groups is a less significant barrier to the efficacy of selection at the level of the group:

Cultural evolution created cooperative, symbolically marked groups. Such environments favoured the evolution of a suite of new social instincts suited to life in such groups, including a psychology which 'expects' life to be structured by moral norms and is designed to learn and internalise such norms; new emotions, such as shame and guilt, which increase the chance the norms are followed; and a psychology which 'expects' the world to be divided into symbolically marked groups. (Richerson and Boyd 2005: 214)

In the specific case of these 'social instincts'—and one can only speculate about their more general views regarding affective states—Richerson and Boyd's view is one where culture affects the emotions in a manner that is primarily *indirect*: emotional instincts evolve under genetic (and not cultural) control, but the fact that they can emerge in this way is a consequence of the establishment of culturally homogeneous groups.

In this chapter I suggest that cultural evolutionists should also explore more direct roles for culture in the shaping of emotions across time. This cultural shaping of emotions is compatible with an evolutionary psychological approach of the sort espoused by Ekman, which claims that distinct human populations share patterns of emotional expression by virtue of their descent from common ancestors. There need be no conflict between evolutionary approaches focused on universal themes in emotional expression and perception, and fine-grained ethnographic approaches focused on very significant differences in the practices woven into those same emotions.

9.2 The Absence of Emotion

Why have the emotions received comparatively little attention in cultural evolutionary research? One answer may lie in the starting point of much of this work. Cultural evolutionists often motivate their basic approach by pointing out, first, that human adaptation typically depends on vast storehouses of complex know-how; and, second, that the transmission and generation of this know-how can plausibly be explained only by appeal to various means by which we learn from interaction with others (Richerson and Boyd 2005, Boyd et al. 2011, Sterelny 2012). Much work in cultural evolutionary theory has, therefore, examined the evolution of various learning biases, which facilitate flows of information across generations. This orients research away from the origin and modification of phenomenally 'hot', or lively, states (fear, disgust, surprise), and concentrates instead on the refinement of states whose phenomenology is apparently colder, or absent altogether (belief, knowledge, and so forth).

Another reason for neglect may lie in a widely held view about the nature of emotions themselves. It has become popular in recent years, especially in evolutionary circles, to think of emotions as embodied states. The view of emotions as 'affect programmes', espoused by Ekman (1992) among others, casts each emotion as a complex syndrome of bodily changes, which may include changes in blood pressure, facial expression, posture, and so forth. Changes to cognitive states might be elements of these syndromes, but they are merely elements. On this view, emotions are partly constituted by bodily reactions to stimuli of various sorts. Damasio (1999) thinks of emotions in a similar way, adding that emotional 'feelings' (which he distinguishes from emotions themselves) are brain states that register the onset of bodily emotions.

Recall from Chapter 3 that when Mesoudi defines culture as 'information', he explains that it 'is intended as a broad term to refer to what social scientists and lay people might call knowledge, beliefs, attitudes, norms, preferences, and skills, all of which may be acquired from other individuals via social transmission and consequently shared across social groups' (2011: 3). I doubt it is accidental that Mesoudi omits emotions from his list, while he does include more purely cognitive-sounding terms such as 'preference' and 'attitude'. Cultural evolutionists have focused on the sorts of states that can be acquired by learning: we can learn skills, beliefs, and preferences through our dealings with others. If emotions, on the other hand, are embodied states, then one might assume they cannot be acquired in these ways: if fear is located in the viscera, then one might worry about how fear could also be learned. That, in turn, might explain why the dominant way of integrating emotions into theories of cultural evolution has been in an indirect manner.

Recall that culture is causally relevant to the promotion of lactose tolerance, but not because lactose tolerance is something we learn from others. Rather, cultural changes—in the form of increased pastoralism—enabled conditions that favoured the evolution of lactose tolerance under natural selection. One might think the only way to integrate emotional states into an account of cultural evolution is in a similar manner: emotional states are not learned—Richerson and Boyd speak of tribal social *instincts*—but their prevalence can be indirectly affected by social learning.

In the remainder of this chapter I sketch some more expansive roles for cultural evolutionary approaches to the emotions. Even if the emotions are embodied, this does not entail that they are unaffected by learning. Of course, this is hardly a new idea: a sort of 'divide and rule' programme has been suggested before, whereby some aspects of the emotions are understood as innate, while others are left to culturally specific rules. I also wish to suggest that the manner in which this division of labour is usually advocated typically takes advantage of problematic distinctions between what is innate and what is acquired, or between nature and culture, or between ontogeny and phylogeny. A cultural evolutionary account that avoids these distinctions is likely to foster greater cooperation between natural and social scientists. Hence, the particular case of the evolution of the emotions is emblematic of the possibilities for a more constructive relationship between these very different approaches to cultural change.

9.3 Learning to be Disgusted

Darwin thought that rational reflection could affect the scope of sympathy, even if sympathy was understood as a state of the body. Far more recent research suggests that other emotions—even when understood as embodied states—can indeed be affected by rational reflection in this manner. Rozin et al. (1997) provide evidence that individuals who are vegetarians on moral grounds (rather than on health grounds) are often initially persuaded that they should not eat meat by reflection on animal rights issues and the pain caused to animals: it is only later that the eating of meat comes to elicit emotional feelings of disgust. This work points to a much broader range of processes whereby activities such as smoking are initially perceived as morally unproblematic, reflection later suggests they may be objectionable, and only after that are they perceived as disgusting. Rozin's vegetarian subjects report being emotionally unable to chew and swallow meat.

Darwin was right, then, to think that rational reflection on the consequences of action could be transformed into altered visceral responses.

If this can happen, we have evidence that cultural changes—changes in the beliefs and preferences of groups of individuals—can directly affect the emotional states of those individuals, even when emotions themselves are understood as embodied states. This also means that, to the extent that cultural evolutionists are interested in the accumulation of storehouses of adaptive cultural information, we can think of embodied emotional states as repositories of this kind of information. Disgust reactions can become tuned to the moral preferences of a community, and they can do so as a result of learning. Disgust reactions, in other words, can orient action in a way that corresponds with the moral norms of a community; in that sense, these reactions contain 'cultural information' in just the same way that Mesoudi thinks skills contain cultural information. They constitute a form of culturally specific know-how.

9.4 Cultural Variation in Basic Emotions

So far the role secured for the direct action of culture on the emotions is rather weak: Rozin's evidence is restricted to disgust, and it also leaves open the thought that the disgust affect programme is unaltered by culture, even if culture can influence the affect programme's elicitors. What we need, then, is evidence that emotions themselves are subject to significant cultural variation, and also that emotions other than disgust can be influenced by culture in this way.

Jesse Prinz (2004a, 2004b) has argued that emotions are at one and the same time embodied and subject to cultural influence, but the weight of this claim turns in part on the rather specific sense in which he thinks of emotions as 'embodied'. In contrast to Damasio, Prinz explicitly denies that emotions should be identified with bodily changes, and he denies that emotions have bodily changes as parts. Prinz says that emotions are *perceptual states*, and hence, states of mind. More specifically, they are perceptions of bodily changes. Fear, for example, is a perception of increased pulse and so forth, but what makes that perceptual state fear, rather than (say) anger, is that it registers danger, rather than offence. That, in turn, means that the perceptual state associated with fear has the function of being caused by dangerous things.

Prinz explains this with an analogy: a smoke detector is not made from smoke and it does not have smoke as one of its parts. Nonetheless, what defines something as a smoke detector is that it registers the presence of smoke. This means that if one modifies a smoke detector, or builds on it in some way, such that it loses the function of registering the presence of smoke, then it is no longer properly thought of as a smoke detector. By defining emotions in terms of the matters of concern they register, Prinz can say that a given perceptual state can change from being of one emotional type to another, when the concerns it registers are altered. Prinz calls emotions 'embodied' because they are perceptions of bodily states, but this does not mean he thinks of emotions as states of the body (2004a).

One might think that Ekman's work undermines Prinz's proposal that emotions are subject to cultural manipulation. Ekman argues for the universality of at least the six 'basic' emotions of fear, disgust, surprise, anger, happiness, and sadness. Ekman allows, of course, that basic emotional affect programmes are subject to culturally mediated (and modifiable) input conditions and rules for display. So, for Ekman, while local rules can dictate what sorts of things are disgusting, and while they can allow for suppression of typical expressions of disgust when such expressions are culturally inappropriate, disgust itself is the same in all cultures.

One reason why Prinz argues that even basic emotions are subject to cultural modification is that, on his view, it is incoherent to say that disgust itself remains unchanged when the culturally specific display rules associated with it are altered. If socialization, reflection, training, and so forth have the result that a group of people systematically suppresses the pattern of bodily changes standardly associated with disgust, this has the result that they perceive a modified set of bodily changes when presented with rotten meat, dead bodies, and so forth. Since emotions for Prinz are just perceptions of bodily changes, this means that emotions themselves change when the bodily changes associated with them change. This is a relatively minor victory over Ekman, because it leaves untouched the notion that six basic affect programmes are unchanging and universal. It tells us only that if Ekman is right that culture can modify display rules, then our account of how emotions are defined should also lead us to say that culture can modify emotions themselves.

A full theory of culturally modifiable emotions would aim at a more substantial result. It would also make room for the idea that core aspects of Ekman's 'affect programmes'—perhaps phenomenological aspects, but also the broader syndromes of facial and bodily changes associated with fear, disgust, and so forth—are also liable to cultural influence. Prinz has also helped with this task in a number of ways (2004a, 2012). Ekman's claim is that all cultures express the six basic emotions in the same ways. But in one of Ekman's early studies of the Fore people, 56 per cent of the Fore thought that what Westerners would regard as a sad face was in fact a face showing what Westerners call anger. Forty-five per cent of the Fore associated what Westerners regarded as a surprised face with fear (Ekman et al. 1969). Following Russell (1994), Prinz suggests that Ekman's results are not as compelling as one might think.

At this point it is important to recognize (as Prinz does, too) that Ekman himself thinks of the basic emotions as 'emotion families' (Ekman 1992). Ekman does not argue that emotions are identical in all cultures; rather, he argues for the existence of what we might think of as a 'clumpy' structuring of emotional space, with six discrete groupings of emotion states within it. Ekman is not committed to the view that fear, for example, is strictly identical in all cultures: indeed, it is consistent with his basic hypothesis that fear varies quite substantially in all respects—including core elements of emotional expression—from one culture to another. Ekman need only deny that these emotion states vary enough to undermine a clustered taxonomy. And since Ekman at least sometimes professes neutrality regarding the developmental processes that produce these affect programmes (as reported, for example, by Camras 1992: 271, fn. 1), his position leaves open the thought that variation *within* emotion families might be produced by various forms of learning and enculturation.

Ekman's thesis of universal emotional clustering has been questioned in recent work by Jack et al. (2012). They suggest that presenting 'Western Caucasians' with a large variety of short computer-generated videos of facial expressions, and asking them to characterize them using the six basic emotion terms, does indeed produce six distinct clusters corresponding to each of these basic emotions. But their work indicates that this clustering does not appear when 'East Asians' perform the same task. In what follows I will assume that Ekman is correct about cross-cultural clustering in the categorization of emotion, but it is worth stressing that even this comparatively mild assertion of clustering is far from trivial.

Ekman denies that the links between basic emotions and their expressions are arbitrary, in the manner that links between words and their referents are arbitrary. Similarity across cultures in terms of fear-like emotions, anger-like emotions, and so forth would be incredible if each culture determined emotional expression by local convention. But it is one thing to deny that emotional expression is *wholly* determined by local cultural convention, another to deny that local cultural traditions can exert *some* significant influences on emotions and their patterns of expression. Ekman may be committed to the first denial, but he is not committed to the second.

It is also important to note the large body of anthropological work that suggests culturally driven variation in emotion. Catherine Lutz's (1988) research has been widely discussed in the context of whether Ekman's basic emotions are truly universal. Lutz suggests considerable cross-cultural variation in emotion terms. It is difficult to give a straightforward English counterpart of the Ifaluk emotion *fago*: it has resonances with compassion, love, and sadness. The Ifaluk emotion of *song* corresponds only loosely to English anger, for while anger can arise in many ways, *song* must arise from a justified cause. One might also look to earlier anthropological work for evidence of cultural variation. Marilyn Strathern (1968) and Andrew Strathern (1968) document an emotion experienced among the people of Mount Hagen, in the Highlands of Papua New Guinea, called *popokl*. *Popokl* is also a form of anger, but it seems more specific than Western anger in terms of how it is provoked, how it is experienced, and how it may be remedied. It is a form of frustration, coupled with a desire for revenge. It is elicited by a failure to acknowledge due social status, or a lack of proper attention: in other words, unlike anger it has a very specific tie to social morality.

The obvious reaction to this anthropological work is to argue that while enculturation may affect emotional *categories*, it does not affect basic emotions themselves (except for when it influences input conditions and display rules). Analogously, it is one thing to ask whether there is cultural variation in the category of 'working memory', another to ask whether there is cultural variation in working memory. A cognitive psychologist is likely to argue that working memory itself is a universal cognitive faculty, while acknowledging that use of the category of 'working memory' is largely restricted to a small cultural group of knowledgeable scientists. And so, we might be entirely content to agree—and we surely must agree—that various forms of learning affect our possession of the concepts of *popokl, song, fago*, anger, and so forth, while denying that emotions themselves are affected by learning.

It is, of course, important to distinguish between concepts and things; between a community's emotion concepts and their emotions; between their psychological concepts and their cognitive make-up. But once this distinction has been drawn, it is equally important to ask how concepts might affect the things they are used to categorize (Hannon 2010). This has been a recurring theme of Ian Hacking's work on 'looping kinds' (e.g. Hacking 1995): Hacking argues, quite plausibly, that various forms of social categorization can affect social reality. Once the category of 'Hispanic' (to use one of his many examples) is established for census purposes, it invites people to describe themselves publicly as Hispanic who might otherwise never have dreamed of doing so, it may bring political unity to a community, it encourages new associations between people, new reflections on shared history, and so forth. The term itself can cause a more cohesive group of people to exist.

Similarly, emotion terms can sanction and thereby bring together otherwise disunified forms of behaviour. The Stratherns explain how, according to the people of Mount Hagen, the ghosts of those who die when *popokl* may be *popokl* themselves. Ghosts who are *popokl* may seek revenge from those who have wronged them, and ghosts may send illness as a result. Somewhat confusingly, ghosts may also send illness to those for whom they feel sympathy: this illness is a way for ghosts to align themselves with those who are *popokl*, which thereby adds further weight to the claims of those who believe themselves to have been wronged. The recognized norms of behaviour for those who are *popokl* can help to re-establish social equilibrium, and the status of individuals as *popokl* or otherwise helps to shape social behaviour by eliciting redress from others. Young children learn about who is and who is not due various forms of respect, they learn to recognize reasonable instances of *popokl* in others, and they do this at the same time as they acquire characteristic *popokl* behaviours when their own roles are slighted. Women who were not properly respected by their husbands are sometimes said to have died 'with *popokl* in their hearts' (M. Strathern 1968). Discussion of *popokl* helps to shape sensitivity to, and worry about, marital neglect, while also provoking anxious reflection on appropriate forms of redress.

Still, one might insist that all these possibilities for looping do not show how basic emotions can change under cultural influence. Mallon and Stich (2000) have argued for a form of compatibility between Ekman's claim that anger is universal, and Lutz's assertion that *song* is culturally specific. They suggest that the Ifaluk have the basic anger emotion just as we do, but they simply label it in a manner that picks it out only when produced by a morally sanctioned cause. On this view, culture has changed only the taxonomy of cognitive processes, not the underlying processes themselves.

Ekman's work is indeed compatible with Lutz's, but I suggest that Mallon and Stich do not go far enough in showing the affinities between Ekman's universalism, and work of a more constructivist style. To return to the Stratherns, even if it turns out that culture influences core aspects of emotional affect programmes, such that *popokl* and anger are properly understood as differing even with respect to these core features, they would still be similar enough with respect to patterns of elicitation and expression for Ekman to retain his claim that (Hagen) *popokl* and (English) anger are members of the same emotion family.

Work by Lutz and the Stratherns shows vividly how the categories of *popokl* and of *song* can influence debate and action in the communities where those terms are used. In Lutz's case such debates concern the morally sanctioned causes of *song*, the treatment of one who acts in a *song* manner without moral justification, and so forth. Of course, even here one might try to stand firm, and argue that anger itself is not subject to alteration, and that *popokl* and *song* are culturally specific categories that happen to pick out instantiations of the universal affect programme of anger in different communities. These categorizations, one might then argue, do not affect anger itself, but they do affect forms of social interaction that issue from anger. This move is certainly not forced on the evolutionist, and it may not be wise. It has been suggested, for example, that alternative forms of enculturation can affect not only terminology and action, but also perception.

There appears to be extensive variation in susceptibility to visual illusions: thanks to Henrich et al.'s (2010) important paper on WEIRD (Western, educated, industrialized, rich, and democratic) research subjects, we have been realerted to a fascinating study from the 1960s indicating that San foragers do not see the Müller-Lyer illusion as an illusion at all, and many other cultures show a far less pronounced response than Americans (Segall et al. 1966). Henrich and colleagues could have mentioned much earlier work on the Müller-Lyer illusion, published in 1901 by the Cambridge anthropologist and physiologist W. H. R. Rivers (1901). Rivers, too, found the responses to the illusion from the undergraduate students he tested in Cambridge to be more extreme than those from the Murray Islanders he tested during his expedition to the Torres Strait. Segall et al. (1966) suggest that vulnerability to the illusion is a result of being raised in 'carpentered' environments: in other words, it is a result of upbringing.

It has also been suggested that colour terms affect abilities to judge differences in colour, although whether this means that colour terms also affect the perceived colours of surfaces remains unclear (Winawer et al. 2007, Masuda 2009). Russian has two different words for blue: *goluboy* and *siniy*, which English speakers would normally refer to as light blue and dark blue respectively. Winawer and colleagues asked subjects to decide which of two coloured squares was the same colour as a third. Their experiment showed that when the two squares also fell into different categories—i.e. when one was *goluboy* and the other *siniy*—Russian speakers made these decisions more quickly than when they fell into the same category. English speakers showed no such advantage. The point is not that English speakers are unable to see differences that are clear to Russian speakers; instead, the experiment suggests that the possession of more finely differentiated colour terms affects how easily one can judge similarity and difference in colour patches. These accentuated abilities of perceptual discrimination are driven, it seems, by a consistent exposure from childhood to a culturally specific set of linguistic terms.

To turn back to the emotions, individuals learn what is to count as an instance of *popokl* in others, they learn how to behave when one is slighted, and all this is bound up with a broader range of modes of behaviour relevant to social morality. If a term like *popokl* works in a similar way to a term like *goluboy*, then we might expect that this form of emotional categorization will also be linked to a collection of heightened discriminatory abilities—to discern swiftly patterns of similarity and difference in the social behaviours of others, and to recognize swiftly the presence of specific dispositions in oneself. It may then turn out that the felt character of the emotion is affected by exposure to a specific set of emotion terms, just as it may turn out—although it still remains conjectural—that Russian speakers perceive colour differences at the *goluboy/siniy* boundary more acutely than do English speakers.

9.5 Nature and Culture: Divisions of Labour

In this penultimate section I very briefly review some existing ways of dividing labour between those aspects of emotion where culture might have influence and those areas where it does not. First, we have already seen reason to be sceptical of Mallon and Stich's suggestion that while basic emotions themselves are universal, the manner in which they are picked out is not. On their view, only the latter is subject to cultural influence. This relies on a dubious premise regarding causal isolation between the manner in which a group categorizes a phenomenon and the phenomenon itself.

Second, we should also be sceptical of Ekman's own occasional insistence on a division of labour between ontogenetic explanation and phylogenetic explanation. For Ekman, the claim that an emotion is 'basic' carries two primary connotations: first, it is universal and, second, its features are to be explained by appeal to evolution. He also notes that because these connotations are logically independent of each other, a 'social constructivist' about emotions might agree to the first while denying the second. We can now see that a social constructivist—of a reasonably sober variety—might agree to both connotations of 'basic'. Since human populations are evolved from common ancestors, then it is possible that this common ancestry explains the similarities between the specific members of a single 'emotion family', while various forms of cultural difference explain why members of a single family diverge to the extent they do. This is just how the evolutionist explains homologous structures in general: shared ancestry explains why forelimbs in dogs, whales, and bats resemble each other, while particular local evolutionary forces, acting on finer-grained taxa, explain their dissimilarities.

Ekman refines his notion of emotion families in the following way:

Each emotion family can be considered to constitute a theme and variations. The theme is composed of the characteristics unique to that family. The variations on that theme are the product of various influences: individual differences in biological constitution; different learning experiences; and differences specific to and reflecting the nature of the particular occasion in which an emotion occurs. Öhman's (1986) description of a multiple-level evolutionary perspective suggests that the themes may be largely the product of our evolution and given genetically, while the variations reflect learning, both species constant and species variable learning experiences. This learning, he maintains (p. 127) is " . . . constrained and shaped by evolution." [. . .] I am proposing that the themes are not simply the most common feature of a basic emotion category, but are the core elements, the product of our evolution, to be found in all instances of an emotion. (1992: 173)

This manner of presenting things invites criticisms that Ekman might easily avoid. We have already seen that Ekman sometimes avows neutrality regarding the question of how emotions develop; however, this passage appears to commit him to a specific view of emotional development. It sketches a picture according to which, in every human culture, there is a core element of basic emotion which is 'given genetically', and various modifications of the core which are affected by culturally local processes. In fact, and to repeat a point made earlier, all that Ekman is committed to is the claim that there are well-defined cross-cultural *clusterings* in emotions, and that the existence of these clusters can be explained by reference to their possession of a common ancestor. That, in turn, does not require that any 'core' is literally common to each basic emotion in every culture: just as homologous limbs in different taxa might differ in significant respects in every one of their features, so the members of emotion families in different cultures might also differ in every respect (Griffiths 1997). Nor is Ekman strictly committed to the claim that when elements of emotion families are held in common, this is because they are 'given genetically'. A model of gradual evolutionary divergence from a common cultural ancestor is compatible with the notion not merely that recent *divergence* in the cultural aspects of developmental environments partly explains emotional divergence, but that more ancient *commonalities* in the cultural aspects of developmental environments partly explain why the development of emotion states remains constant across cultures.

This last claim is, of course, wholly speculative: I have no empirically established account to offer of the ways in which culturally ubiquitous features of emotions develop. But recall Heyes's (2012a) work, suggesting that very robust, pan-cultural features of our species may rely for their development in important ways on forms of socialization. The ability of humans to imitate others is just the sort of widely distributed trait that one might naïvely think must be innate, or 'genetically specified'. Heyes argues that this capacity is learned, and she consequently questions inferences from cross-cultural ubiquity to innateness.

To turn to comparative work on the emotions, William Mason's developmental studies of infant rhesus macaques that are deprived of social contact suggests that while they are able to form facial expressions characteristic of fear (grimace), friendliness (lip-smacking), and threat, they are unable to use these expressive patterns to manage their relationships with conspecifics (Griffiths 2004). As Mason puts it:

one of the consequences of experience is that the rhesus monkey becomes not only more adept at using expressive movements to manage others but does so in a manner that suggests the behaviors are used in ways that are more socially expedient than expressive. (Mason 1985: 146)

Mason's dysfunctional monkeys are deprived of the social feedback experiences that normally result in emotions that function in 'regulation and control of social life' (ibid.). So, here too, social learning is important for the functionality of emotions, even when these functional elements of emotion are typical of development in the wild, and even when they have evolutionary significance. Sarah Hrdy, in her work on the evolution of parenting, has also stressed the manner in which the development of nurturing responses towards children has relied throughout hominin history on suitable social interactions. Some of her reasoning draws on psychological work exploring the contingency of human capacities for collaboration on the provision of care in infancy by 'responsive others' (Hrdy 2009: 291). The upshot, she argues, is that emotional responses to others that are—or at least have been—characteristic of our species are malleable in the face of changed circumstances of upbringing: 'why wouldn't novel modes of childrearing continue to shape not just child development, but human nature?' (2009: 292). It would be a mistake, then, to assume that the distinction between features that are culturally universal and culturally specific lines up with a distinction between the innate and the acquired, or with a distinction between traits that can be explained in evolutionary terms and traits that can instead be explained in terms of cultural influence.

All of this points to an intriguing compatibility between 'practice theory' in the social sciences more generally and cultural evolutionary approaches to the emotions. If we note the recent origins of practice theory in the work of Pierre Bourdieu, and the more distal origins of the movement in writings of Marcel Mauss, then a few themes emerge. First is the Maussian notion that (in Bloch's paraphrase) 'the bodily, the mental and the psychological are simultaneously present in social relations' (Bloch 2012, 149). The cultural evolutionist need not deny any of these claims in the context of emotion: emotions can be at one and the same time embodied, mental, psychological, and social. Even Mason's macaques possess emotional states which allow them to manage the behaviour of others, and which are partially constituted by aspects of facial movement. Second is the stress from Bourdieu not merely on embodiment, but also on the concept of 'habitus'. At least part of what Bourdieu appears to have in mind when he deploys this notion is a belief that the development of modes of individual action requires situation within a suitably structured social environment.

As I read him, this is another way of saying that the more or less reliable development of, for example, various emotional states requires life within an environment that is suitably structured, and suitably structured by virtue of the patterned actions (and interactions) of others. Once again, we can see this sort of claim borne out experimentally by Mason's macaques: when deprived of a suitable background of structured social interactions, the proper development of an ability to act through the appropriate formation of facial expressions is hampered. Social anthropologists who have been attracted to forms of practice theory have typically been sceptical of evolutionary approaches to culture, but there are hopes for an empirically grounded union of these approaches.

The view of the emotions sketched in this chapter shows how cultural evolutionary theory can be freed from conceptual shackles that might otherwise impede that theory's development. First, there are good reasons to think that cultural processes can affect the development of the emotions, even emotions usually regarded as basic. Second, this means that we should not propose a strict contrast between constructivist and evolutionary accounts of the emotions. Just as Darwin thought, it is possible for the emotions to be at one and the same time products of long phylogenetic histories, while also being influenced in their development throughout long periods of these histories by social interaction of various kinds. Third, this image of emotions as subject to modification by social interaction, and more specifically by learning, does not depend on assuming a cognitivist picture of emotions: even if we think of emotions as embodied, we can still make room for cultural influence over such states. Fourth, this in turn means that there is no barrier to cultural evolutionary theorists including emotional states explicitly in their lists of states that contain 'cultural information'. Fifth, cultural evolutionists need not think that all cultural information is 'stored in human brains' (Richerson and Boyd 2005: 5); some may instead be contained within embodied emotional states.

9.6 The Descent of Culture

Dan Sperber opens his seminal case for a naturalistic approach to culture by telling us that 'a spectre haunts the social sciences, the spectre of a natural science of the social' (1996: vi). If the arguments of this book are correct, the social sciences have little to fear, and something to gain, from these other-worldly visitations. I have suggested that there is little credible prospect for radically reorganizing the social sciences around a central commitment to a selectionist approach to cultural change. I have also argued that it is implausible to think that reflection on the very broad demands of ancestral environments will offer significant heuristic leverage when we come to ask how people think and interact today.

What evolutionary approaches to culture have to offer is more modest. They bring a set of useful tools to students of culture. Phylogenetic models help us to put constraints on the historical narratives offered for the development of material and linguistic culture. Mathematical populational models help us to explore with precision how groups of individuals act in aggregation. These tools will help to illuminate some of the questions addressed by social scientists, but not all of those questions. When they work at their best, evolutionary models will not supplant work in developmental and social psychology, ethnography, history, and so forth: instead, their assumptions will be deferential to the deliveries of these more traditional approaches.

We have seen that hostility to cultural evolutionary theory sometimes derives from the impression that it is committed to untenable distinctions between what is universal and what is cultural, between what is given by evolved nature and what is given by local culture, between what is genetically specified and what is learned, between culture in the brain and biology in the body. We have also seen that an evolutionary approach to cultural change need be committed to none of them. In understanding all this, we can allay the suspicions of cultural and social anthropologists, and show how the natural and social sciences might ultimately be knitted back together.

References

Almeida, C., S. Lorena, C. Pavan, H. Akasaka, and M. Mesquita (2012) 'Beneficial Effects of Long-Term Consumption of a Probiotic Combination of *Lactobacillus casei* Shirota and *Bifidobacterium breve* Yakult May Persist after Suspension of Therapy in Lactose-Intolerant Patients', *Nutrition in Practice*, 27: 247–51.

Ariew, A. (1996) 'Innateness and Canalization', *Philosophy of Science* (Proceedings), 63: S19–27.

Ariew, A. (1999) 'Innateness is Canalization', in V. Hardcastle (ed.), *Where Biology Meets Psychology: Philosophical Essays*, Cambridge, MA: MIT Press, pp. 117–38.

Asch, S. E. (1955) 'Opinions and Social Pressure', *Scientific American*, 193: 31–55.

Atran, S. (2001) 'The Trouble with Memes', *Human Nature*, 12: 351–81.

Avital, E., and E. Jablonka (2000) *Animal Traditions: Behavioural Inheritance in Evolution*, Cambridge: Cambridge University Press.

Barbrook, A., C. Howe, N. Blake, and P. Robinson (1998) 'The Phylogeny of *The Canterbury Tales*', *Nature*, 394: 839.

Basalla, G. (1988) *The Evolution of Technology*, Cambridge: Cambridge University Press.

Bentley, R. A., C. Lipo, H. Herzog, and M. Hahn (2007) 'Regular Rates of Popular Culture Change Reflect Random Copying', *Evolution and Human Behavior*, 28: 151–8.

Bergstrom, C., and M. Rosvall (2011) 'The Transmission Sense of Information', *Biology and Philosophy*, 26: 159–76.

Blackmore, S. J. (2000) *The Meme Machine*, Oxford: Oxford University Press.

Bloch, M. (1998) *How We Think They Think: Anthropological Approaches to Cognition, Memory, and Literacy*, Boulder, CO: Westview Press.

Bloch, M. (2012) *Anthropology and the Cognitive Challenge*, Cambridge: Cambridge University Press.

Boehm, D. (1999) *Hierarchy in the Forest: The Evolution of Egalitarian Behavior*, Cambridge, MA: Harvard University Press.

Borgerhoff Mulder, M. (2001) 'Using Phylogenetically Based Comparative Methods in Anthropology: More Questions than Answers', *Evolutionary Anthropology*, 10: 99–111.

Bourdieu, P. (1977) *Outline of a Theory of Practice*, Cambridge: Cambridge University Press.

Boyd, R., and P. Richerson (1985) *Culture and the Evolutionary Process*, Chicago: University of Chicago Press.

Boyd, R., and P. Richerson (2000) 'Memes: Universal Acid or a Better Mousetrap?', in R. Aunger (ed.), *Darwinizing Culture: The Status of Memetics as a Science*, Oxford: Oxford University Press, pp. 143–62.

Boyd, R., P. Richerson, and J. Henrich (2011) 'The Cultural Niche: Why Social Learning is Essential for Human Adaptation', *PNAS*, 108: 10918–25.

Boyer, P. (2001) *Religion Explained: The Evolutionary Origins of Religious Thought*, New York: Basic Books.

Buller, D. J. (2005) *Adapting Minds: Evolutionary Psychology and the Persistent Quest for Human Nature*, Cambridge, MA: MIT Press.

Buss, L. (1987) *The Evolution of Individuality*, Princeton, NJ: Princeton University Press.

Campbell, D. (1974) 'Evolutionary Epistemology', in P. A. Schilpp (ed.), *The Philosophy of Karl Popper*, La Salle, IL: Open Court, pp. 412–63.

Camras, L. (1992) 'Expressive Development and Basic Emotions', *Cognition and Emotion*, 6: 269–83.

Cavalli-Sforza, L., and M. Feldman (1981) *Cultural Transmission and Evolution: A Quantitative Approach*. Princeton, NJ: Princeton University Press.

Champagne, F., and M. Meaney (2007) 'Transgenerational Effects of Social Environment on Variations in Maternal Care and Behavioral Response to Novelty', *Behavioral Neuroscience*, 121: 1353–63.

Chomsky, N., and M. Foucault (2006) *The Chomsky–Foucault Debate on Human Nature*, New York: The New Press.

Claidière, N., and D. Sperber (2007) 'Commentary: The Role of Attraction in Cultural Evolution', *Journal of Cognition and Culture*, 7: 89–111.

Claidière, N., and A. Whiten (2012) 'Integrating the Study of Conformity and Culture in Humans and Nonhuman Animals', *Psychological Bulletin*, 138: 126–45.

Claidière, N., M. Bowler, and A. Whiten (2012) 'Evidence for Weak or Linear Conformity but Not for Hyper-Conformity in an Everyday Social Learning Context', *PLoS ONE*, 7: e30970.

Claidière, N., T. Scott-Phillips, and D. Sperber (2014) 'How Darwinian Is Cultural Evolution?', *Philosophical Transactions of the Royal Society B*, 369: 20130368.

Clark, A., and D. Chalmers (1998) 'The Extended Mind', *Analysis*, 58: 7–19.

Cosmides, L., J. Tooby, and J. Barkow (1992) 'Introduction: Evolutionary Psychology and Conceptual Integration', in J. Barkow, L. Cosmides and J. Tooby (eds) *The Adapted Mind*, New York: Oxford University Press, pp. 3–15.

Cosmides, L., and J. Tooby (1997) 'Evolutionary Psychology: A Primer'. Available online: <http://www.cep.ucsb.edu/primer.html>. Accessed 26 February 2014.

Daly, M., and M. Wilson (1988) *Homicide*, New York: De Gruyter.

Damasio, A. (1999) *The Feeling of What Happens*, New York: Harcourt, Brace and Co.

Danchin, É., L. Giraldeau, T. Valone, and R. Wagner (2004) 'Public Information: From Nosy Neighbours to Cultural Evolution', *Science*, 305: 487–91.

Danchin, É., A. Charmantier, F. Champagne, A. Mesoudi, B. Pujol, and S. Blanchet (2011) 'Beyond DNA: Integrating Inclusive Inheritance into an Extended Theory of Evolution', *Nature Reviews Genetics*, 12: 475–86.

Darwin, C. (1859) *The Origin of Species*, London: John Murray.

Darwin, C. (1862) Letter to Asa Gray, 4 September, Darwin Correspondence Letter 3710. Available online: <http://www.darwinproject.ac.uk/entry-3710>. Accessed 20 February 2014.

Darwin, C. (1871) *The Descent of Man and Selection in Relation to Sex*, London: John Murray.

Darwin, C. (1872) *The Expression of the Emotions in Man and the Animals*, London: John Murray.

Dawkins, R. (1989) *The Selfish Gene*, 2nd edn, Oxford: Oxford University Press.

Dawkins, R. (2008) 'Why Darwin Matters', *The Guardian*, 9 February. Available online: <http://www.guardian.co.uk/science/2008/feb/09/darwin.dawkins1>. Accessed 20 February 2014.

Dennett, D. C. (1996) *Darwin's Dangerous Idea: Evolution and the Meanings of Life*, London: Penguin.

Dennett, D. C. (2001) 'In Darwin's Wake, Where Am I?', *Proceedings and Addresses of the American Philosophical Association*, 75: 11–30.

Descola, P. (2013) *Beyond Nature and Culture*, Chicago: Chicago University Press.

Devitt, M. (2008) 'Resurrecting Biological Essentialism', *Philosophy of Science*, 75: 344–82.

Devitt, M. (2010) 'Species Have (Partly) Intrinsic Essences', *Philosophy of Science*, 77: 648–61.

Dixon, T. (2003) *From Passions to Emotions*, Cambridge: Cambridge University Press.

Dupré, J. (1993) *The Disorder of Things: Metaphysical Foundations for the Disunity of Science*, Cambridge, MA: Harvard University Press.

Dupré, J. (2003) *Human Nature and the Limits of Science*, Oxford: Oxford University Press.

Dupré, J. (2012) *Processes of Life*, Oxford: Oxford University Press.

Durkheim, É. (2006) 'Rules for the Explanation of Social Facts', in H. Moore and T. Sanders (eds), *Anthropology in Theory*, Oxford: Blackwell, pp. 54–63.

Edelman, G. (1987) *Neural Darwinism*, New York: Basic Books.

Efferson, C., R. Lalive, P. Richerson, R. McElreath, and M. Lubell (2008) 'Conformists and Mavericks: The Empirics of Frequency-Dependent Cultural Transmission', *Evolution and Human Behavior*, 29: 56–64.

Ekman, P. (1992) 'An Argument for Basic Emotions', *Cognition and Emotion*, 6: 169–200.

Ekman, P., E. Sorenson, and W. Friesen (1969) 'Pan-Cultural Elements in Facial Displays of Emotions', *Science*, 164: 86–8.

Engelberg, J., and L. Boyarsky, (1979) 'The Noncybernetic Nature of Ecosystems', *American Naturalist*, 114: 317–24.

Enquist, M., P. Strimling, K. Eriksson, K. Laland, and J. Sjostrand (2010) 'One Cultural Parent Makes No Culture', *Animal Behaviour*, 79: 1353–62.

Ereshefsky, M., and M. Matthen (2005) 'Taxonomy, Polymorphism and History: An Introduction to Population Structure Theory', *Philosophy of Science*, 72: 1–21.

Eriksson, K., M. Enquist, and S. Ghirlanda (2007) 'Critical Points in Current Theory of Conformist Social Learning', *Journal of Evolutionary Psychology*, 5: 67–87.

Fodor, J. (1983) *The Modularity of Mind*, Cambridge, MA: MIT Press.

Fortunato, L., and R. Mace (2009) 'Testing Functional Hypotheses about Cross-Cultural Variation: A Maximum-Likelihood Comparative Analysis of Indo-European Marriage Practices', in S. Shennan (ed.), *Pattern and Process in Cultural Evolution*, Berkeley: University of California Press, pp. 235–50.

Fracchia, J., and R. C. Lewontin (1999) 'Does Culture Evolve?', *History and Theory*, 38: 52–78.

Fracchia, J., and R. C. Lewontin (2005) 'The Price of Metaphor', *History and Theory*, 44: 14–29.

Frank, R. (1988) *Passions within Reason: The Strategic Role of the Emotions*, New York: Norton.

Geertz, C. (1973) 'Thick Description: Toward an Interpretive Theory of Culture', in C. Geertz (ed.), *The Interpretation of Cultures*, New York: Basic Books, pp. 3–30.

Godfrey-Smith, P. (1996) *Complexity and the Function of Mind in Nature*, Cambridge: Cambridge University Press.

Godfrey-Smith, P. (2000) 'The Replicator in Retrospect', *Biology and Philosophy*, 15: 403–23.

Godfrey-Smith, P. (2001) 'On the Status and Explanatory Structure of Developmental Systems Theory', in R. D. Gray, P. E. Griffiths, and S. Oyama (eds), *Cycles of Contingency: Developmental Systems and Evolution*, Cambridge, MA: MIT Press, pp. 283–98.

Godfrey-Smith, P. (2009) *Darwinian Populations and Natural Selection*, Oxford: Oxford University Press.

Godfrey-Smith, P. (2012) 'Darwinism and Cultural Change', *Philosophical Transactions of the Royal Society B*, 367: 2160–70.

Goldstein, L., and W. Plaut (1955) 'Direct Evidence for Nuclear Synthesis of Cytoplasmic Ribose Nucleic Acid', *Proceedings of the National Academy of Sciences of the United States of America*, 41: 874–80.

Gould, S. J. (1988) *An Urchin in the Storm*, New York: Norton.

Gould, S. J., and R. C. Lewontin (1979) 'The Spandrels of San Marco and the Panglossian Paradigm: A Critique of the Adaptationist Programme', *Proceedings of the Royal Society of London B*, 205: 581–98.

Gray, R. D. (1992) 'Death of the Gene: Developmental Systems Strike Back', in P. E. Griffiths (ed.), *Trees of Life: Essays on the Philosophy of Biology*, Dordrecht: Kluwer, pp. 165–209.

Gray, R. D., S. J. Greenhill, and R. M. Ross (2007) 'The Pleasures and Perils of Darwinizing Culture (with Phylogenies)', *Biological Theory*, 2: 360–75.

Griesemer, J., M. Haber, G. Yamashita, and L. Gannett (2005) 'Critical Notice: *Cycles of Contingency—Developmental Systems and Evolution*', *Biology and Philosophy*, 20: 517–44.

Griffiths, P. E. (1996) 'The Historical Turn in the Study of Adaptation', *British Journal for the Philosophy of Science*, 47: 511–32.

Griffiths, P. E. (1997) *What Emotions Really Are: The Problem of Psychological Categories*, Chicago: University of Chicago Press.

Griffiths, P. E. (2001) 'Genetic Information: A Metaphor in Search of a Theory', *Philosophy of Science*, 68: 394–412.

Griffiths, P. E. (2002) 'What Is Innateness?', *Monist*, 85: 70–85.

Griffiths, P. E. (2004) 'Towards a Machiavellian Theory of Emotional Appraisal', in D. Evans and P. Cruse (eds), *Emotion, Evolution and Rationality*, Oxford: Oxford University Press, pp. 89–105.

Griffiths, P. E. (2009) 'Reconstructing Human Nature', *Arts*, 31: 30–57.

Griffiths, P. E., and R. D. Gray (1994) 'Developmental Systems and Evolutionary Explanation', *The Journal of Philosophy*, 91: 277–304.

Griffiths, P. E., and R. D. Gray (1997) 'Replicator II: Judgement Day', *Biology and Philosophy*, 12: 471–92.

Griffiths, P. E., and R. D. Gray (2001) 'Darwinism and Developmental Systems', in R. D. Gray, P. E. Griffiths, and S. Oyama (eds), *Cycles of Contingency: Developmental Systems and Evolution*, Cambridge, MA: MIT Press, pp. 195–218.

Hacking, I. (1995) 'The Looping Effects of Human Kinds', in D. Sperber, D. Premack, and A. James Premack (eds), *Causal Cognition: A Multidisciplinary Debate*, Oxford: Clarendon Press, pp. 351–94.

Hannon, E. (2010) 'The Nurture of Nature: Biology, Psychology and Culture', unpublished PhD thesis, University of Durham.

Harris, P. L., and K. H. Corriveau (2011) 'Young Children's Selective Trust in Informants', *Philosophical Transactions of the Royal Society B*, 366: 1179–87.

Henare, A. J. M., M. Holbraad, and S. Wastell (2007) *Thinking through Things: Theorising Artefacts Ethnographically*, London: Routledge.

Henrich, J. (2001) 'Cultural Transmission and the Diffusion of Innovations: Adoption Dynamics Indicate that Biased Cultural Transmission is the

Predominate [*sic*] Force in Behavioral Change', *American Anthropologist*, 103: 992–1013.

Henrich, J., and R. Boyd (1998) 'The Evolution of Conformist Transmission and the Emergence of Between-Group Differences', *Evolution and Human Behavior*, 19: 215–41.

Henrich, J., and R. Boyd (2002) 'On Modelling Culture and Cognition: Why Cultural Evolution Does Not Require Replication of Representations', *Culture and Cognition*, 2: 87–112.

Henrich, J., and J. Broesch (2011) 'On the Nature of Cultural Transmission Networks: Evidence from Fijian Villages for Adaptive Learning Biases', *Philosophical Transactions of the Royal Society B*, 366: 1139–48.

Henrich, J., and F. Gil-White (2001) 'The Evolution of Prestige: Freely Conferred Deference as a Mechanism for Enhancing the Benefits of Cultural Transmission', *Evolution and Human Behavior*, 22: 165–96.

Henrich, J., R. Boyd, and P. Richerson (2012) 'The Puzzle of Monogamous Marriage', *Philosophical Transactions of the Royal Society B*, 367: 657–69.

Henrich, J., S. Heine, and A. Norenzayan (2010) 'The Weirdest People in the World?', *Behavioral and Brain Sciences*, 33: 61–135.

Heyes, C. (1994) 'Social Learning in Animals: Categories and Mechanisms', *Biological Reviews*, 69: 207–31.

Heyes, C. (2001) 'Causes and Consequences of Imitation', *Trends in Cognitive Sciences*, 5: 253–61.

Heyes, C. (2010) 'Where Do Mirror Neurons Come From?', *Neuroscience and Biobehavioral Reviews*, 34: 575–83.

Heyes, C. (2012a) 'Grist and Mills: On the Cultural Origins of Cultural Learning', *Philosophical Transactions of the Royal Society B*, 367: 2181–91.

Heyes, C. (2012b) 'What's Social about Social Learning?', *Journal of Comparative Psychology*, 126: 193–202.

Heyes, C. (2012c) 'New Thinking: The Evolution of Human Cognition', *Philosophical Transactions of the Royal Society B* 367: 2091–6.

Hobaiter, C., T. Poisot, K. Zuberbühler, W. Hoppitt, and T. Gruber (2014) 'Social Network Analysis Shows Direct Evidence for Social Transmission of Tool Use in Wild Chimpanzees', *PLoS Biology*, 12: e1001960.

Hodgson, G., and T. Knudsen (2010) *Darwin's Conjecture: The Search for General Principles of Social and Economic Evolution*, Chicago: University of Chicago Press.

Holden, C., and R. Mace (1997) 'Phylogenetic Analysis of the Evolution of Lactose Digestion in Adults', *Human Biology*, 69: 605–28.

Horner, V., D. Proctor, K. Bonnie, A. Whiten, and F. de Waal (2010) 'Prestige Affects Cultural Learning in Chimpanzees', *PLoS One*, 5: e10625.

Houkes, W. (2012) 'Population Thinking and Natural Inheritance in Dual Inheritance Theory', *Biology and Philosophy*, 27: 401–17.

Hrdy, S. (2009) *Mothers and Others: The Evolutionary Origins of Mutual Under-standing*, Cambridge, MA: Harvard University Press.

Hull, D. (1986) 'Human Nature', *PSA: Proceedings of the Biennial Meeting of the Philosophy of Science Association*, 2: 3–13.

Hull, D. (1988) *Science as a Process*, Chicago: University of Chicago Press.

Ingold, T. (1995) 'People Like Us: The Concept of the Anatomically Modern Human', *Cultural Dynamics*, 7: 187–214.

Ingold, T. (2007) 'The Trouble with "Evolutionary Biology"', *Anthropology Today*, 23: 13–17.

Ingold, T. (2013) 'Prospect', in T. Ingold and G. Palsson (eds), *Biosocial Becomings: Integrating Social and Biological Anthropology*, Cambridge: Cambridge University Press, pp. 1–21.

Ingram, C., C. Mulcare, Y. Itan, M. Thomas, and D. Swallow (2009) 'Lactose Digestion and the Evolutionary Genetics of Lactase Persistence', *Human Genetics*, 124: 579–91.

Itan, Y., B. Jones, C. Ingram, D. Swallow, and M. Thomas (2010) 'A Worldwide Correlation of Lactase Persistence Phenotypes and Genotypes', *BMC Evolutionary Biology*, 10: 36.

Itan, Y., A. Powell, M. Beaumont, J. Burger, and M. Thomas (2009) 'The Origins of Lactase Persistence in Europe', *PLoS Computational Biology*, 5: e1000491.

Jablonka, E. (2001) 'The Systems of Inheritance', in R. D. Gray, P. E. Griffiths, and S. Oyama (eds), *Cycles of Contingency: Developmental Systems and Evolution*, Cambridge, MA: MIT Press, pp. 99–116.

Jablonka, E. (2002) 'Information: Its Interpretation, its Inheritance, and its Sharing', *Philosophy of Science*, 69: 578–605.

Jablonka, E., and M. J. Lamb (1998) 'Epigenetic Inheritance in Evolution', *Journal of Evolutionary Biology*, 11: 159–83.

Jablonka, E., and M. J. Lamb (2005) *Evolution in Four Dimensions: Genetic, Epigenetic, Behavioral, and Symbolic Variation in the History of Life*, Cambridge, MA: MIT Press.

Jablonka, E., and E. Szathmary (1995) 'The Evolution of Information Storage and Heredity', *Trends in Ecology and Evolution*, 10: 206–11.

Jack, R., O. Garrod, H. Yui, R. Caldara, and P. Schyns (2012) 'Facial Expressions of Emotion Are Not Culturally Universal', *Proceedings of the National Academy of Sciences of the United States of America*, 109: 7241–4.

Kitcher, P. (1985) *Vaulting Ambition: Sociobiology and the Quest for Human Nature*, Cambridge, MA: MIT Press.

Knauft, B. (1991) 'Violence and Sociality in Human Evolution', *Current Anthropology*, 32: 391–428.

Knudsen, S. (2005) 'Communicating Novel and Conventional Scientific Metaphors: A Study of the Development of the Metaphor of Genetic Code', *Public Understanding of Science*, 14: 373–92.

Krebs, J., and N. Davies (1997) 'Introduction to Part Two', in J. R. Krebs and N. B. Davies (eds), *Behavioural Ecology: An Evolutionary Approach*, 4th ed., Oxford: Blackwell Science, pp. 15–18.

Kroeber, A. (1917) 'The Superorganic', *American Anthropologist*, 19: 163–213.

Kronfeldner, M. (2007) 'Is Cultural Evolution Lamarckian?', *Biology and Philosophy*, 22: 493–512.

Kuper, A. (2000a) 'If Memes Are the Answer, What Is the Question?', in R. Aunger (ed.), *Darwinizing Culture: The Status of Memetics as a Science*, Oxford: Oxford University Press, pp. 175–88.

Kuper, A. (2000b) *Culture: The Anthropologists' Account*, Cambridge, MA: Harvard University Press.

Laland, K. N., and G. R. Brown (2002) *Sense and Nonsense: Evolutionary Perspectives on Human Behaviour*, Oxford: Oxford University Press.

Laland, K. N., J. Odling-Smee, and M. Feldman (2001) 'Niche Construction, Biological Evolution, and Cultural Change', *Behavioral and Brain Sciences*, 23: 131–46.

Laland, K. N., K. Sterelny, J. Odling-Smee, W. Hoppitt, and T. Uller (2011) 'Cause and Effect in Biology Revisited: Is Mayr's Proximate–Ultimate Dichotomy Still Useful?', *Science*, 334: 1512–16.

Layton, R. (1997) *An Introduction to Theory in Anthropology*, Cambridge: Cambridge University Press.

Layton, R. (2010) 'Why Social Scientists Don't Like Darwin and What Can Be Done about It', *Journal of Evolutionary Psychology*, 8: 139–52.

Levins, R. (1970) 'Complexity', in C. H. Waddington (ed.), *Towards a Theoretical Biology: Drafts*, Edinburgh: University of Edinburgh Press, pp. 67–86.

Levinthal, C. (1959) 'Coding Aspects of Protein Synthesis', *Reviews of Modern Physics*, 31: 249–55.

Levy, A. (2011) 'Information in Biology: A Fictionalist Account', *Noûs*, 45: 640–57.

Lewens, T. (2002a) 'Darwinnovation!', *Studies in History and Philosophy of Science*, 33: 199–207.

Lewens, T. (2002b) 'Adaptationism and Engineering', *Biology and Philosophy*, 17: 1–31.

Lewens, T. (2004) *Organisms and Artifacts: Design in Nature and Elsewhere*, Cambridge, MA: MIT Press.

Lewens, T. (2007) *Darwin*, London: Routledge.

Lewens, T. (2009a) 'Innovation and Population', in U. Krohs and P. Kroes (eds), *Functions in Biological and Artificial Worlds: Comparative Philosophical Perspectives*, Cambridge, MA: MIT Press, pp. 243–58.

Lewens, T. (2009b) 'What is Wrong with Typological Thinking?', *Philosophy of Science*, 76: 355–71.

Lewens, T. (2009c) 'Seven Types of Adaptationism', *Biology and Philosophy*, 24: 161–82.

Lewens, T. (2010a) 'Natural Selection Then and Now', *Biological Reviews*, 85: 829–35.

Lewens, T. (2010b) 'The Natures of Selection', *British Journal for the Philosophy of Science*, 61: 313–33.

Lewens, T. (2012a) 'Pheneticism Reconsidered', *Biology and Philosophy*, 27: 159–77.

Lewens, T. (2012b) 'Species, Essence and Explanation', *Studies in History and Philosophy of Biological and Biomedical Sciences*, 43: 751–7.

Lewens, T. (2012c) 'Human Nature: The Very Idea', *Philosophy and Technology*, 25: 459–74.

Lewens, T. (2012d) 'The Darwinian View of Culture', *Biology and Philosophy*, 27: 745–53.

Lewens, T. (2013) 'From *Bricolage* to Biobricks™: Synthetic Biology and Rational Design', *Studies in History and Philosophy of Biological and Biomedical Sciences*, 44: 641–8.

Lewens, T. (2014) 'Review of *The Evolved Apprentice*', *British Journal for the Philosophy of Science*, 65: 185–9.

Lewens (2015) 'Backwards in Retrospect', *Philosophical Studies*, 172: 813–21.

Lewis, D. (1969) *Convention: A Philosophical Study*, Cambridge, MA: Harvard University Press.

Lewontin, R. C. (1983) 'The Organism as the Subject and Object of Evolution', *Scientia*, 118: 63–82.

Lewontin, R. C. (2005) 'The Wars over Evolution', *New York Review of Books*, 20 October. Available online: <http://www.nybooks.com/articles/archives/2005/oct/20/the-wars-over-evolution/>. Accessed 20 February 2014.

Lipton, P. (2004) *Inference to the Best Explanation*, London: Routledge.

Lloyd, E. (1999) 'Evolutionary Psychology: The Burdens of Proof', *Biology and Philosophy*, 14: 211–33.

Lukes, S. (2005) *Power: A Radical View*, 2nd edn, Basingstoke: Palgrave Macmillan.

Lutz, C. (1988) *Unnatural Emotions: Everyday Sentiments on a Micronesian Atoll and the Challenge to Western Theory*, Chicago: University of Chicago Press.

Mace, R. (2010) 'Update to Holden and Mace's "Phylogenetic Analysis of the Evolution of Lactose Digestion in Adults"', *Human Biology*, 81: 621–4.

Mace, R., and C. J. Holden (2005) 'A Phylogenetic Approach to Cultural Evolution', *Trends in Ecology and Evolution* 20: 116–21.

Mace, R., and F. Jordan (2011) 'Macro-Evolutionary Studies of Cultural Diversity: A Review of Empirical Studies of Cultural Transmission and Cultural Adaptation', *Philosophical Transactions of the Royal Society B*, 366: 402–11.

McElreath, R., M. Lubell, P. Richerson, T. Waring, W. Baum, E. Edsten, C. Efferson, and B. Paciotti (2005) 'Applying Evolutionary Models to the Laboratory Study of Social Learning', *Evolution and Human Behavior*, 26: 483–508.

Machery, E. (2008) 'A Plea for Human Nature', *Philosophical Psychology*, 21: 321–9.

Machery, E. (2012) 'Reconceptualising Human Nature: Response to Lewens', *Philosophy and Technology*, 25: 475–8.

McKitrick, M. (1993) 'Phylogenetic Constraint in Evolutionary Theory: Has It Any Explanatory Power?', *Annual Review of Ecology and Systematics*, 24: 307–30.

Mallon, R., and S. Stich (2000) 'The Odd Couple', *Philosophy of Science*, 67: 133–54.

Mameli, M. (2004) 'Nongenetic Selection and Nongenetic Inheritance', *British Journal for the Philosophy of Science*, 55: 35–71.

Mameli, M. (2008a) 'Understanding Culture: A Commentary on Richerson and Boyd's *Not by Genes Alone*', *Biology and Philosophy*, 23: 269–81.

Mameli, M. (2008b) 'On Innateness: The Clutter Hypothesis and the Cluster Hypothesis', *Journal of Philosophy*, 105: 719–36.

Mameli, M., and P. Bateson (2006) 'Innateness and the Sciences', *Biology and Philosophy*, 21: 155–88.

Mameli, M., and P. Bateson (2011) 'An Evaluation of the Concept of Innateness', *Philosophical Transactions of the Royal Society B*, 366: 436–43.

Mason, W. (1985) 'Experiential Influences on the Development of Expressive Behaviours in Rhesus Monkeys', in G. Zivin (ed.), *The Development of Expressive Behaviour*, New York: Academic Press, pp. 117–52.

Masuda, T. (2009) 'Cultural Effects on Visual Perception', in E. Goldstein (ed.), *Encyclopedia of Perception*, Volume 1, Thousand Oaks, CA: Sage Publications, pp. 339–43.

Matthews, L., J. Tehrani, F. Jordan, M. Collard, and C. Nunn (2011) 'Testing for Divergent Transmission Histories among Cultural Characters: A Study Using Bayesian Phylogenetic Methods and Iranian Tribal Textile Data', *PLoS One*, 6: e14810.

Maynard Smith, J. (2000) 'The Concept of Information in Biology', *Philosophy of Science*, 67: 177–94.

Maynard Smith, J., and Szathmary, E. (1995) *The Major Transitions in Evolution*, Oxford: Oxford University Press.

Mesoudi, A. (2011) *Cultural Evolution: How Darwinian Theory Can Explain Human Culture and Synthesize the Social Sciences*, Chicago: University of Chicago Press.

Mesoudi, A., and S. Lycett (2009) 'Random Copying, Frequency-Dependent Copying and Culture Change', *Evolution and Human Behavior*, 30: 41–8.

Mesoudi, A., D. Veldhuis, and R. Foley (2010) 'Why Aren't the Social Sciences Darwinian?', *Journal of Evolutionary Psychology*, 8: 93–104.

Mesoudi, A., A. Whiten, and K. Laland (2004) 'Perspective: Is Human Cultural Evolution Darwinian? Evidence Reviewed from the Perspective of *The Origin of Species*', *Evolution*, 58: 1–11.

Mesoudi, A., A. Whiten, and K. Laland (2006) 'Towards a Unified Science of Cultural Evolution', *Behavioral and Brain Sciences*, 29: 329–47.

Mesoudi, A., A. Whiten, and K. Laland (2007) 'Science, Evolution and Cultural Anthropology: A Response to Ingold', *Anthropology Today*, 23: 18.

Mithen, S. (2000) 'Mind, Brain, and Material Culture: An Archaeological Perspective', in P. Carruthers and A. Chamberlain (eds), *Evolution and the Human Mind: Modularity, Language and Meta-Cognition*, Cambridge: Cambridge University Press, pp. 207–17.

Morange, M. (2010) 'How Evolutionary Biology Presently Pervades Cell and Molecular Biology', *Journal for General Philosophy of Science*, 41: 113–20.

Morgan, T., and K. Laland (2012) 'The Biological Bases of Conformity', *Frontiers in Neuroscience*, 6: 87.

Morgan, T., L. Rendell, M. Ehn, W. Hoppitt, and K. Laland (2012) 'The Evolutionary Basis of Human Social Learning', *Proceedings of the Royal Society B*, 279: 653–62.

Morin, O. (2011) *Comment les traditions naissent et meurent: la transmission culturelle*, Paris: Odile Jacob.

Moss, L. (2003) *What Genes Can't Do*, Cambridge, MA: MIT Press.

O'Brien, M., and R. L. Lyman (2002) 'Evolutionary Archeology: Current Status and Future Prospects', *Evolutionary Anthropology*, 11: 26–36.

O'Brien, M., B. Buchanan, M. Collard, and M. Boulanger (2012) 'Cultural Cladistics and the Early Prehistory of North America', in P. Pontarotti (ed.), *Evolutionary Biology: Mechanisms and Trends*, Berlin: Springer, pp. 23–42.

O'Brien, M., R. L. Lyman, A. Mesoudi, and T. L. VanPool (2010) 'Cultural Traits as Units of Analysis', *Philosophical Transactions of the Royal Society B*, 365: 3797–806.

Odenbaugh, J., and A. Alexandrova (2011) 'Buyer Beware: Robustness Analyses in Economics and Biology', *Biology and Philosophy*, 26: 757–71.

Odling-Smee, J., K. Laland, and M. Feldman (2003) *Niche Construction: The Neglected Process in Evolution*, Princeton, NJ: Princeton University Press.

Öhman, A. (1986) 'Face the Beast and Face the Fear: Animal and Social Fears as Prototypes for Evolutionary Analyses of Emotion', *Psychophysiology*, 23: 123–45.

Oyama, S. (1985) *The Ontogeny of Information: Developmental Systems and Evolution*, Durham, NC: Duke University Press.

Oyama, S. (2000) *The Ontogeny of Information: Developmental Systems and Evolution*, 2nd edn, Durham, NC: Duke University Press.

Perry, G., and R. Mace (2010) 'The Lack of Acceptance of Evolutionary Approaches to Human Behaviour', *Journal of Evolutionary Psychology*, 8: 105–25.

Pinker, S. (2002) *The Blank Slate: The Modern Denial of Human Nature*, London: Allen Lane.

Popper, K. (1962) *Conjectures and Refutations*, London: Routledge.

Prinz, J. (2004a) 'Which Emotions are Basic?', in D. Evans and P. Cruse (eds), *Emotion, Evolution and Rationality*, Oxford: Oxford University Press, pp. 69–88.

Prinz, J. (2004b) *Gut Feelings*, Cambridge, MA: MIT Press.

Prinz, J. (2012) *Beyond Human Nature: How Culture and Experience Shape Our Lives*, London: Allen Lane.

Ramsey, G. (2012) 'How Human Nature Can Inform Human Enhancement: A Commentary on Tim Lewens's *Human Nature: The Very Idea*', *Philosophy and Technology*, 25: 479–83.

Ramsey, G. (2013a) 'Human Nature in a Post-Essentialist World', *Philosophy of Science*, 80: 983–93.

Ramsey, G. (2013b) 'Culture in Humans and Other Animals', *Biology and Philosophy*, 28: 457–79.

Ray, E., and C. Heyes (2011) 'Imitation in Infancy: The Wealth of the Stimulus', *Developmental Science*, 14: 92–105.

Richardson, R. (2007) *Evolutionary Psychology as Maladapted Psychology*, Cambridge, MA: MIT Press.

Richerson, P., and R. Boyd (2005) *Not By Genes Alone: How Culture Transformed Human Evolution*, Chicago: University of Chicago Press.

Richerson, P., and J. Henrich (2012) 'Tribal Social Instincts and the Cultural Evolution of Institutions to Solve Collective Action Problems', *Cliodynamics*, 3: 38–80.

Risjord, M. (2007) 'Ethnography and Culture', in S. Turner and M. Risjord (eds), *Philosophy of Anthropology and Sociology*, Amsterdam: Elsevier, pp. 399–428.

Rivers, W. H. R. (1901) *Reports of the Cambridge Anthropological Expedition to Torres Straits: Physiology and Psychology*, Volume 2, Cambridge: Cambridge University Press.

Rose, R. (2012) 'Semantic Information in Genetics: A Metaphorical Reading', unpublished MPhil essay, University of Cambridge.

Rozin, P., M. Markwith, and C. Stoess (1997) 'Moralization and Becoming a Vegetarian: The Transformation of Preferences into Values and the Recruitment of Disgust', *Psychological Science*, 8: 67–73.

Rublack, U. (2004) *Reformation Europe*, Cambridge: Cambridge University Press.

Russell, J. (1994) 'Is There Universal Recognition of Emotion from Facial Expression?', *Psychological Bulletin*, 95: 102–41.

Ryle, G. (1971) 'The Thinking of Thoughts: What is "Le Penseur" Doing?', in G. Ryle (ed.), *Collected Essays 1928–68*, Volume 2, London: Hutchinson.

Salganik, M., P. Dodds, and D. Watts (2006) 'Experimental Study of Inequality and Unpredictability in an Artificial Cultural Market', *Science*, 311: 854–6.

Samuels, R. (2012) 'Science and Human Nature', *Royal Institute of Philosophy Supplement*, 70: 1–28.

Schrödinger, E. (1944) *What Is Life?* Cambridge: Cambridge University Press.

Segall, M., D. T. Campbell, and M. Herskovits (1966) *The Influence of Culture on Visual Perception*, Indianapolis, IN: Bobbs-Merrill.

Shea, N. (2007) 'Representation in the Genome and in Other Inheritance Systems', *Biology and Philosophy*, 22: 313–31.

Shea, N. (2012) 'New Thinking, Innateness and Inherited Representation', *Philosophical Transactions of the Royal Society B*, 367: 2234–44.

Shea, N. (2013) 'Inherited Representations are Read in Development', *British Journal for the Philosophy of Science*, 64: 1–31.

Shuster, S. (1987) 'Alternative Reproductive Behaviors: Three Discrete Male Morphs in *Paracerceis Sculpta*', *Journal of Crustacean Biology*, 7: 318–27.

Sinervo, B., and C. M. Lively (1996) 'The Rock-Paper-Scissors Game and the Evolution of Alternative Male Strategies', *Nature* 380: 240–3.

Skyrms, B. (1996) *Evolution of the Social Contract*, Cambridge: Cambridge University Press.

Smith, R. (2007) *Being Human: Historical Knowledge and the Creation of Human Nature*, New York: Columbia University Press.

Sober, E. (1992) 'Models of Cultural Evolution', in P. E. Griffiths (ed.), *Trees of Life: Essays on the Philosophy of Biology*, Dordrecht: Kluwer, pp. 17–40.

Sober, E. (1993) *Philosophy of Biology*, Boulder, CO: Westview.

Sober, E. (2008) *Evidence and Evolution: The Logic Behind the Science*, Cambridge: Cambridge University Press.

Sober, E. (2009) 'Did Darwin Write the *Origin* Backwards?', *Proceedings of the National Academy of Sciences of the United States of America*, 106: 10048–55.

Sperber, D. (1996) *Explaining Culture: A Naturalistic Approach*, Oxford: Blackwell.

Sperber, D. (2000) 'An Objection to the Memetic Approach to Culture', in R. Aunger (ed.), *Darwinizing Culture: The Status of Memetics as a Science*, Oxford: Oxford University Press, pp. 163–74.

Sperber, D. (2006) 'Why a Deep Understanding of Cultural Evolution is Incompatible with Shallow Psychology', in N. C. Enfield and S. C. Levinson (eds), *Roots of Human Sociality: Culture, Cognition and Interaction*, Oxford: Berg, pp. 431–49.

Sperber, D., and N. Claidière (2006) 'Why Modeling Cultural Evolution Is Still Such a Challenge', *Biological Theory*, 1: 20–2.

Sperber, D., and N. Claidière (2008) 'Defining and Explaining Culture (comments on Richerson and Boyd, *Not by Genes Alone*)', *Biology and Philosophy*, 23: 283–92.

Sterelny, K. (2001) 'Niche Construction, Developmental Systems, and the Extended Replicators', in R. D. Gray, P. E. Griffiths, and S. Oyama (eds), *Cycles of Contingency: Developmental Systems and Evolution*, Cambridge, MA: MIT Press, pp. 333–50.

Sterelny, K. (2003) *Thought in a Hostile World*, Oxford: Blackwell.

Sterelny, K. (2006) 'Memes Revisited', *The British Journal for the Philosophy of Science*, 57: 145–65.

Sterelny, K. (2010) 'Moral Nativism: A Sceptical Response', *Mind and Language*, 25: 279–97.

Sterelny, K. (2012) *The Evolved Apprentice: How Evolution Made Humans Unique*, Cambridge, MA: MIT Press.

Sterelny, K., and P. E. Griffiths (1999) *Sex and Death*, Chicago: University of Chicago Press.

Sterelny, K., K. Smith, and M. Dickison (1996) 'The Extended Replicator', *Biology and Philosophy*, 11: 377–403.

Strathern, A. (1968) 'Sickness and Frustration: Variation in Two New Guinea Highlands Societies', *Mankind*, 6: 545–51.

Strathern, M. (1968) 'Popokl: The Question of Morality', *Mankind*, 6: 553–62.

Swallow, D. (2003) 'Genetics of Lactase Persistence and Lactose Intolerance', *Annual Review of Genetics*, 37: 197–219.

Thompson, A. (1997) 'An Evolved Circuit, Intrinsic in Silicon, Entwined with Physics', in T. Higuchi, M. Iwata, and L. Weixin (eds), *Evolvable Systems: From Biology to Hardware*, Berlin: Springer, pp. 390–405.

Thompson, E. (2007) *Mind in Life*, Cambridge, MA: Harvard University Press.

Tooby, J., and L. Cosmides (1990) 'On the Universality of Human Nature and the Uniqueness of the Individual: The Role of Genetics and Adaptation', *Journal of Personality*, 1: 17–67.

Tooby, J., and L. Cosmides (1992) 'The Psychological Foundations of Culture', in J. Barkow, L. Cosmides, and J. Tooby (eds), *The Adapted Mind: Evolutionary Psychology and the Generation of Culture*, Oxford: Oxford University Press, pp. 19–136.

Toren, C. (2012) 'Anthropology and Psychology', in R. Fardon, O. Harris, T. H. J. Marchand, C. Shore, V. Stang, R. A. Wilson, and M. Nuttall (eds), *The Sage Handbook of Social Anthropology*, Volume 1, Los Angles: Sage, pp. 27–41.

Turchin, P. (2009a) 'Long-Term Population Cycles in Human Societies', *Annals of the New York Academy of Sciences*, 1162: 1–17.

Turchin, P. (2009b) 'A Theory for Formation of Large Empires', *Journal of Global History*, 4: 191–217.

Vygotsky, L. (1986) *Thought and Language*, Cambridge, MA: MIT Press.

Wakano, J., and K. Aoki (2007) 'Do Social Learning and Conformist Bias Coevolve? Henrich and Boyd Revisited', *Theoretical Population Biology*, 72: 504–12.

Weaver, I., N. Cervoni, F. Champagne, A. D'Alessio, S. Sharma, J. Secki, S. Dymov, M. Szyf, and M. Meaney (2004) 'Epigenetic Programming by Maternal Behavior', *Nature Neuroscience*, 7: 847–54.

Weisberg, M., and K. Reisman (2008) 'The Robust Volterra Principle', *Philosophy of Science*, 75: 106–31.

Wimsatt, W. (1999) 'Genes, Memes, and Cultural Heredity', *Biology and Philosophy*, 14: 279–310.

Wimsatt, W. (2006) 'Aggregate, Composed, and Evolved Systems: Reductionistic Heuristics as a Means to More Holistic Theories', *Biology and Philosophy*, 21: 667–702.

Winawer, J., N. Witthoft, M. Frank, L. Wu, A. Wade, and L. Boroditsky (2007) 'Russian Blues Reveal Effects of Language on Color Discrimination', *Proceedings of the National Academy of Sciences of the United States of America*, 104: 7780–5.

Witteveen, J. (2013) 'Rethinking "Typological" versus "Population" Thinking: A Historical and Philosophical Re-Assessment of a Troubled Dichotomy', unpublished PhD dissertation, University of Cambridge.

Ziman, J. (2000) *Technological Innovation as an Evolutionary Process*, Cambridge: Cambridge University Press.

Index